PRAISE FOR

THE AGE
OF
WOOD

"A stunning book on the incalculable debt humanity owes to wood . . . Roland Ennos's knowledge of all things arboreal is vast and intricate. He is a professor of biology at the University of Hull and the author of several books, among them the Natural History Museum's official guide to trees. But [*The Age of Wood*] is something different—nothing less than a complete reinterpretation of human history and prehistory, and it is written with enormous verve and pinpoint clarity. . . . No review can match the richness of Ennos's book. There are chapters or sections on coal and charcoal, pottery kilns, modern wooden buildings, techniques of melting and smelting metals, the history of shipbuilding, wind and watermills, deforestation, and much else. . . . I felt like cheering."

—*The Sunday Times* (UK)

"Ennos, a professor at the University of Hull in England and a specialist in the mechanical properties of trees, shares his insatiable curiosity with us. He applies his sharp eye for details, and he does so entertainingly."

—*The Washington Post*

"Ennos's special love and concern is for things made from trees. . . . The principles of every significant technology, from tree-felling and carpentry to shipbuilding and papermaking, are described with a precise, almost mesmerizing detail."

—*The New York Times Book Review*

"A lively history of biology, mechanics, and culture that stretches back 60 million years . . . A specialist in the mechanics of wood, Ennos has a fierce love for his topic."

—*Nature*

"Nearly the whole of human history deserves a different title: the Age of Wood."

—*The New Republic*

"An excellent, thorough history in an age of our increasingly fraught relationships with natural resources."

—*Kirkus Reviews* (starred review)

"This engaging natural history will draw in fans of Mark Kurlansky's *Cod* and Vince Beiser's *The World in a Grain*. It does a fantastic job of elevating humble wood to its rightful place alongside stone, bronze, and iron as a key resource in leading humanity to its dazzling achievements."

—*Library Journal*

"This expansive history will give readers a newfound appreciation for one of the world's most ubiquitous yet overlooked materials."

—*Publishers Weekly*

"Smart and surprising, Ennos's inquiry proves that there is much we still need to learn about wood and how it has shaped our past and present."

—*Booklist*

"This fascinating book is an eye-opening history of wood. . . . From how trees, and our interactions with trees, have shaped ecosystems, to how wood itself has been incorporated into societies, to how wood functions as a material, it gives a rundown like no other."

—*BookMarks*

THE AGE OF WOOD

OUR MOST USEFUL MATERIAL *AND* *THE* CONSTRUCTION *OF* CIVILIZATION

ROLAND ENNOS

SCRIBNER

New York London Toronto Sydney New Delhi

Scribner
An Imprint of Simon & Schuster, Inc.
1230 Avenue of the Americas
New York, NY 10020

First Scribner trade paperback edition December 2021

SCRIBNER and design are registered trademarks of The Gale Group, Inc.,
used under license by Simon & Schuster, Inc., the publisher of this work.

For information about special discounts for bulk purchases,
please contact Simon & Schuster Special Sales at 1-866-506-1949
or business@simonandschuster.com.

The Simon & Schuster Speakers Bureau can bring authors to your live event.
For more information or to book an event, contact the Simon & Schuster Speakers
Bureau at 1-866-248-3049 or visit our website at www.simonspeakers.com.

Manufactured in the United States of America

3   5   7   9   10   8   6   4   2

Library of Congress Cataloging-in-Publication Data has been applied for.

ISBN 978-1-9821-1473-2
ISBN 978-1-9821-1474-9 (pbk)
ISBN 978-1-9821-1475-6 (ebook)

Illustrations on pages 15, 16, 32, 88, 122, 183 © Roland Ennos; page 90 © Duncan Slater.
Insert photograph credits: 1. The Natural History Museum/Alamy; 2. Clearview/
Alamy; 3. Tegel W. et al. 2012. "Early Neolithic Water Wells Reveal the World's
Oldest Wood Architecture." *Plos One* 7 (12): e51374; 4. Philip Bishop/Alamy;
5. Andrej Peunik/© Museum and Galleries of Ljubljana; 6. Stefan Lippmann/
Oneworld Picture/Alamy; 7. Werner Forman Archive/Egyptian Museum, Cairo/
Heritage Images; 8. Jorge Tutor/Alamy; 9. With special permission from the City
of Bayeux; 10. Andreas Werth/Alamy; 11. Boris Baggs/Arcaid Images/Alamy;
12. Markus Lange/Alamy; 13. Adam Woolfitt/Alamy; 14. Zhang Shuo/TAO Images
Limited/Alamy; 15. Greg Balfour Evans/Alamy; 16. Courtesy of the Metropolitan
Museum of Art, Purchase, Bequest of John L. Cadwalader, by exchange, and Peek
Family Foundation Gift, 2016; 17. Courtesy of the Metropolitan Museum of Art,
Purchase, Robert Alonzo Lehman Bequest, 2012; 18. National Anthropological
Archives/Smithsonian Institution; 19. Library of Congress; 20. Library of Congress;
21. Antony Nettle/Alamy; 22.eye35.pix/Alamy; 23. NinaRundsveen/Courtesy of
Wikimedia Commons.

To Robin Wootton,
the best of mentors

Contents

CONTENTS

PART 3:

WOOD IN THE INDUSTRIAL ERA

PART 4:

FACING THE CONSEQUENCES

The Road to Nowhere

Many years ago, toward the end of an arduous walking tour of the French Pyrenees, my brother and I stumbled across an engineering feat that had helped change the course of human history and shape the modern world. As we made our way down from the peaks to the village of Etsaut, the route took us from alpine meadows to the conifer forests of the Vallée d'Aspe. The path, which had been broad and easy to follow, suddenly changed. As the river valley continued to drop, the path maintained its level, but only by cutting into the walls of an almost-sheer rock face. Soon we were walking along a narrow ledge perched precariously six hundred feet above the trees and foaming river in the Gorge d'Enfer below. The path continued like this for almost a mile before the gorge finally opened out, and we descended down to the level of the river and once again felt safe. Only then did a sign helpfully tell us that we had navigated the Chemin de la Mâture. Why had such a spectacular path been built in the middle of nowhere? And what was *mâture*?

The answer lies in the rivalry that developed in the eighteenth century between the two emerging superpowers of the Western world, France and Britain, and provides just one of the more striking examples of the way wood has helped shape the human story. With the two nations vying for power and influence over their developing colonies and territories in the Caribbean and North America, an arms race started as they built up their navies. Both nations strove to build bigger and more heavily armed ships of the line, capable of acting as firing platforms for up to a hundred huge cannons, which could batter other ships and shore defenses into submission. But both countries came up against the same problem; how could they access enough trees to build their ships? The problem was not the lack of wood itself. France in particular had large areas of forest, which covered around 30 percent of the country. The problem was the lack of trees tall and straight enough to make the 100-to-120-foot masts of the ships. Most forests in Europe were already being managed, and it was becoming harder to find areas of primary forest where tall trees could still be found. For France the answer lay in the wilds of the Pyrenees, where stands of huge fir trees still stood. The engineer Paul-Marie Leroy put forward his plan to extract trees from the previously inaccessible Vallée d'Aspe by cutting a daring path through the edge of the cliff. The path was completed in 1772 and named the Chemin de la Mâture (literally, the Mast Road). Soon masts and other timbers were being hauled down the new path, before being rafted down to the sea. France's supply problems were fixed, at least temporarily.

In Britain the problem of obtaining masts was even more acute. The country had a tree cover below 10 percent, and its forests had long before been put under management. Few conifers grew there, and no trees tall and straight enough to be made into ships' masts. Even by the sixteenth century, Britain had been forced to obtain almost all its masts from the countries adjoining the Baltic Sea. The problem was that the fleets of its northern rivals, Holland and Sweden, were always threatening to cut off this supply, and in any case tall trees were becoming scarcer and more expensive. Britain turned to its American colonies, where the old-growth forests of New England contained huge, straight-trunked eastern white pine trees in seemingly limitless numbers. From the mid-seventeenth century onward these trees, which could grow up to 230 feet tall with a diameter of over five feet, became the tree of choice for the British navy; Samuel Pepys, the naval administrator, mentions the trade several times in his famous diary, rejoicing on December 3, 1666, when a convoy carrying masts managed to evade a Dutch blockade:

> There is also the very good news come of four New England ships come home safe to Falmouth with masts for the King; which is a blessing mighty unexpected, and without which, if for nothing else, we must have failed the next year. But God be praised for thus much good fortune, and send us the continuance of his favour in other things!

Unfortunately, in seeking to secure their supply of masts, the British government made a series of policy blunders that were to have disastrous consequences. They had difficulty buying tree trunks on the open market because the colonists preferred to saw them up for timber; this was after all a much easier way of processing them, considering their huge size, rather than hauling the unwieldy trunks for miles down to navigable rivers. The British could have bought up areas of forest and managed them themselves, but instead, in 1691 they implemented what was known as the King's Broad Arrow policy. White pine trees above twenty-four inches in trunk diameter were marked with three strokes of a hatchet in the shape of an upward-pointing arrow and were deemed to be crown property. Unfortunately, this policy soon proved to be wildly unpopular and totally unenforceable. Colonists continued to fell the huge trees and cut them into boards twenty-three inches wide or less, to dispose of the evidence. Indeed wide floorboards became highly fashionable, as a mark of an independent spirit. The British responded by rewriting the protection act to prohibit the felling of all white pine trees over twelve inches in diameter. However, because trees were protected only if they were not "growing within any township or the bounds, lines and limits thereof," the people of New Hampshire and Massachusetts promptly realigned their borders so that the provinces were divided almost entirely into townships. Many rural colonists just ignored the rules, pleaded ignorance of them, or deliberately targeted the marked trees because of their obvious value. The surveyors general of His Majesty's Woods, employing few men and needing to cover tens of thousands of square miles, were almost powerless to stop the depredations of the colonists, and the local authorities were unwilling to enforce an unpopular law. The situation reached a crisis in 1772, exactly when the Chemin de la Mâture was being completed, with the event known as the Pine Tree Riot.

The event was precipitated when sawmill owners from Weare, New Hampshire, refused to pay a fine for sawing up large white pines, and Benjamin Whiting, sheriff of Hillsborough County, and his deputy, John Quigley, were sent to South Weare with a warrant to arrest the leader of the mill owners, Ebenezer Mudgett. However, before they could complete their task, Mudgett led a force of twenty to forty men to assault them at their lodgings, the Pine Tree Tavern. Their faces blackened with soot, the rioters gave the sheriff one lash with a tree switch for every tree being contested, cut off the ears and shaved the manes and tails off Whiting's and Quigley's horses and forced the two men to ride out of town through a gauntlet of jeering townspeople. Eight of the perpetrators were later punished, but their fines, twenty shillings each, were light, an indication of the weakness of British authority.

News of the riot spread around New England and became a major inspiration for the much more famous Boston Tea Party in December 1773. The Pine Tree Flag even became a symbol of colonial resistance, being one of those used by the revolutionaries in the ensuing War of Independence. Designed by George Washington's secretary Colonel Joseph Reed, it was flown atop the masts of the colonial warships.

The start of the Revolutionary War cut off the supply of masts for the Royal Navy from New England. The British were forced to use smaller trees from the Baltic for their masts, and had to clamp together several trunks with iron hoops to construct "made masts." This arrangement was at best unsatisfactory, and many British ships spent most of the ensuing war out of action in port with broken masts. To make matters worse, the colonists started to sell their pines to the French, who had opportunistically sided with the rebels. The French defeated the British in important naval conflicts—such as the Battle of Grenada in 1779, the most disastrous British naval defeat since Beachy Head in 1690—while British naval actions against the colonists themselves proved indecisive. Without Britain's usual naval superiority, America prevailed and became independent in 1783. What would become the world's most powerful nation had been born. Britain would soon regain its naval supremacy, managing to replace its supplies of masts by using trees from its other dominions, Canada and eventually New Zealand, but the world would never again be the same. Thus is a turning point in geopolitics glimpsed in a path hewn out of a cliff in the Pyrenees.

Considering its historical importance, it is astonishing that the Great Mast Crisis is not better known. All schoolchildren are taught about the Boston Tea Party, even in Britain; none are taught about the Pine Tree Riot. But this is not an isolated instance; accounts of human evolution, prehistory, and history routinely ignore the role played by wood. For instance, anthropologists wax lyrical about the developments of stone tools, and the intellectual and motor skills needed to shape them, while brushing aside the importance of the digging sticks, spears, and bows and arrows with which early humans actually obtained their food. Archaeologists downplay the role wood fires played in enabling modern humans to cook their food and smelt metals. Technologists ignore the way in which new metal tools facilitated better woodworking to develop the groundbreaking new technologies of wheels and plank ships. And architectural historians ignore the crucial role of wood in roofing medieval cathedrals, insulating country houses, and underpinning whole cities.

When I stumbled across the Chemin de la Mâture thirty-five years ago, I too was largely ignorant of the importance of wood. I knew about its anatomy, its mechanical properties, and some of its structural uses. However, only when I turned to research the mechanics of root anchorage in plants and landed a permanent post in academia did I start to learn more about wood. One of the great benefits of being an academic (or it used to be) is that it gives you the opportunity to find out about a wide variety of topics, through your own research and teaching, and through discussions with your colleagues in (now sadly defunct) tearooms. In my case, I started to find out more about biomechanics by supervising a wide range of student projects. I set bright young students to study subjects such as the mechanical design of our own bodies, the mechanics of wood and trees, and latterly the benefits of urban forests. I wrote a book about trees and started to learn more about the uses of wood and the relationship between human beings and trees. My teaching also led me to think more about the relationship that our relatives the apes have with trees, and to learn about exciting new research that was uncovering the ways in which apes make and use a variety of wooden tools. I was lucky enough to become involved with researchers who studied how apes move through the canopy and build wooden nests. And I started to think about how early humans could have made effective woodworking tools and shaped their spears and ax handles.

All these discoveries tied in with my happy memories of visits I had made from childhood onward to a wide range of wood-related attractions: local archaeological museums with their rows of ax heads and reconstructions of the life of "early man"; Scandinavian open-air museums, filled with wooden farmhouses, water mills, windmills, and stave churches; Viking longboats; the roofs of Gothic churches and cathedrals, medieval barns and castles; and Palladian country houses. It became clear to me that wood has actually played a central role in our history. It is the one material that has provided continuity in our long evolutionary and cultural story, from apes moving about the forest, through spear-throwing hunter-gatherers and ax-wielding farmers to roof-building carpenters and paper-reading scholars. And knowing something about the properties of wood and the growth of trees, I started to work out why this was the case. The foundations of our relationship with wood lie in its remarkable properties. As an all-round structural material it is unmatched. It is lighter than water, yet weight for weight is as stiff, strong, and tough as steel and can resist both being stretched and compressed. It is easy to shape, as it readily splits along the grain, and is soft enough to carve, especially when green. It can be found in pieces large enough to hold up houses, yet can be cut up into tools as small as a toothpick. It can last for centuries if it is kept permanently dry or wet, yet it can also be burned to keep us warm, to cook our food, and drive a wide range of industrial processes. With all these advantages, the central role of wood in the human story was not just explicable, but inevitable.

So it is time to reassess the role of wood. This book is a new interpretation of our evolution, prehistory, and history, based on our relationship with this most versatile material. I hope to show that looking at the world in this fresh wood-centered way, what an academic might call lignocentric, can help us make far more sense of who we are, where we have come from, and where we are going.

Above all I hope to encourage the reader to look at the world in a way that is unhindered by the conventional wisdom that the story of humanity is defined by our relationship with three materials: stone, bronze, and iron. It refutes the common assumption that wood is little more than an obsolete relic from our distant past. I hope it will show that for the vast majority of our time on this planet we have lived in an age dominated by this most versatile material, and that in many ways we still do. And that for the benefit of the environment and our own physical and psychological health, we need to return to the Age of Wood.

PART 1

WOOD AND
HUMAN EVOLUTION

CHAPTER 1

Our Arboreal Inheritance

In the Western world, we tend to stand aloof from nature and regard ourselves as superior beings who are quite separate from the animal world. Indeed in the Bible creation story, God said, "Let us make man in our image, after our likeness: and let them have dominion over the fish of the sea, and over the fowl of the air, and over the cattle, and over all the earth, and over every creeping thing that creepeth upon the earth." It was natural for biblical writers to emphasize our uniqueness when the only mammals they saw—ungulates such as sheep, goats, camels, and horses; carnivores such as dogs and cats; and rodents such as mice and rats—all walked around on four legs and had limbs ending in hooves or claws. Things look quite different in tropical countries, where people live alongside monkeys and apes. There they stress our similarities to primates and our continuity with nature. Those primitive primates, the galagos of West Africa, are commonly known as bush babies, for instance, while in the Malay language, orangutan literally means "man of the forest." Many religions have monkey gods, from the monkey king Sun Wukong in China, the monkey god Hanumen of

India, and the ancient Egyptians' baboon god Babi. The barrier between ourselves and other animals is most porous in Borneo, where the Dayaks have a legend that orangutans could talk if they wanted to but prefer to remain silent as they do not want to be forced to work: a mark, surely, of the profoundest wisdom!

The big divide between primates and other mammals stems from the ways in which primates are adapted to a life in trees. And despite our now being terrestrial animals, we resemble other primates because we have retained most of these arboreal adaptations. Surprisingly, we were preadapted to our life on the ground by the evolution of our relatives' bodies and brains to live in the forest canopy: in a world made of wood.

Most of the physical changes that primates underwent occurred in the first 10 million years or so of their evolution, shortly after the first primates—small, shrewlike mammals— colonized the rain forests that sprang up 60 million years ago, following the demise of the dinosaurs. We know that because we share those adaptations with those most adorable creatures, the bush babies, which resemble nothing more than miniature, furry humans. Though they are similar to us in so many ways, bush babies are only distant relatives. Fossil evidence and DNA analysis show that their lineage split from ours around 50 million years ago. Yet they share with us many key derived characteristics: binocular vision, with the eyes both pointing forward; an upright body posture; differentiation of the limbs between hind legs and feet for locomotion, and arms and hands for gripping; and soft pads and nails on the tips of their digits, instead of claws. We usually think of these characteristics as being human adaptations, but they actually first evolved to help primates live in trees.

If you think about it, a tree is a tricky place in which to live.

The wooden structure has a complex branching shape, with a vertical trunk that bifurcates successively into more horizontal and thinner boughs, branches, and twigs, structures that eventually end in the productive parts of the tree, the leaves. Having binocular vision helps primates judge distances and move about more quickly and more safely around the canopy. The upright body and grasping arms of primates, meanwhile, allow them to grip on to the trunk and clamber up and down the tree; but it is among the narrow branches and twigs at the ends of the canopy that the modifications of the hands and fingers come into play.

The sharp claws of modern-day squirrels, tree shrews, and woodpeckers are good at finding purchase in the bark of a tree's trunk and branches but ill-suited to hold on to narrow twigs. These animals cannot therefore easily reach the ends of the canopy where most of the leaves and fruit are located. The early primates overcame this difficulty by evolving a key suite of features that are shared by all their descendants, and which have since been vital to our success as toolmakers: gripping hands (and in most other primates, feet) equipped with soft digital pads that are covered in prints and backed not by claws, but nails.

Despite our fingers being at the ends of our hands, few scientists have thought thought much about their design and why we have soft finger pads. Physics textbooks tell us that harder, rougher surfaces should provide the best grip because the projections interlock with those on the substrate. However, this is plainly untrue if the substrate you are trying to grip is smooth; think how easily hobnailed boots slip across smooth rocks. Counterintuitively, the key to getting a better grip on a smooth surface is not to use a hard material such as a claw, but a soft one, such as skin. This increases friction because a soft material deforms to the shape of the other surface, so a large area is in contact, maximizing the interatomic forces between the two surfaces. The softer the material, the more it can deform and the larger the contact area.

To improve our grip, we could cover our finger pads with a biological rubber such as elastin, but this would wear away too fast. The solution evolved by primates is more ingenious: we use a soft internal fluid within our finger pads and surround it by a stiffer lining—producing a structure rather like a partially deflated car tire. Beneath the tips of our fingers are pads of fat, which deform easily to allow a large surface area of the more rigid surrounding skin to make contact. You can see how effective this arrangement is by gripping a wineglass and looking through the other side; you'll be able to see a large area of contact. This arrangement gives us an excellent grip on hard surfaces such as glass, ten times as good as that of hard hooves or claws—explaining why we remain sure-footed on smooth concrete and tiles, whereas horses are prone to slip in their stables, and panicking dogs often scrabble about on the kitchen floor without being able to move off.

Our finger pads and the palms of our hands and feet are also decorated by another characteristic feature: the pattern of ridges known as fingerprints. On smooth materials such as glass, this makes our grip worse, since it reduces the area of contact, just as grooved tires in racing cars have poorer grip in the dry than slicks. However, fingerprints do give some important advantages. They can improve our grip in the wet (just like grooved tires) since they can channel away the surface film of water, and also on rough surfaces, such as branches, since the ridges interlock with ones in the bark. And the skin ridges where our touch receptors are located can magnify strains and so improve the sensitivity of our fingers. Finally, the alternation of strong ridges with flexible troughs in the skin allow it to deform smoothly when we grip an object, preventing blistering. Skin ridges are so useful for improving grip that the totally unrelated koala bears of Australia have evolved similar ridges on their finger pads, while New World monkeys also have prints on the pads of their prehensile tails.

Since their finger pads allowed primates to hold on to narrow branches and twigs so well, they no longer had any need for claws; instead these were flattened into self-trimming nails, which act as a hard backing to the pads, just as the rims of wheels act as a backing to car tires, and help us pick up and manipulate tiny objects. We can even use the tips of our nails as tools themselves, for scratching or prizing small objects apart.

By 50 million years ago, therefore, primates had already made the physical changes that we have since found so helpful to master life on the ground, but the early primates were still very different from us. They were tiny, having a body weight well under a pound; in contrast, modern monkeys typically weigh in between 2 and 35 pounds, and great apes including ourselves weigh anything from 90 to 265 pounds. And they were nowhere near as intelligent. Bush babies have brains that are only slightly larger than those of other mammals of the same size, and only 47 percent of their brains is composed of the neocortex, the gray matter on the surface of the cerebral hemispheres that deals with higher-level thinking. This is large compared with the brain of an insectivore such as a hedgehog, in which this figure is around 18 percent, but small compared with 70 percent for a macaque, 76 percent for a chimpanzee, and 80 percent for us humans. It is starting to become clear that these three characteristics, body size, neocortical size, and intelligence, are actually linked—primates got smarter as they got bigger—and that these changes are related to their arboreal lifestyle.

Primatologists are learning that the reason monkeys increased in size as they evolved was related to changes in their diets. Bush babies and their relatives the lorises are insectivores; they eat insects and other invertebrates, which are hard to find, hard to catch, and rather small. Insects provide enough energy to support a bush baby. However, a larger creature would be no better at finding, catching, and eating insects, but the amount of energy it would have to expend moving about to do so would be much greater. A large insectivorous monkey would not be able to catch enough food to fuel its body. But there are other things that primates could eat in the canopy instead; they could become vegetarian and eat leaves or fruit. Depending on which of these two foods they eat, modern monkeys have evolved rather different body adaptations and have very different intellects.

Leaves are extremely plentiful and easy to find in a rain forest, where all the trees are evergreens, but they make rather unsatisfactory food. Leaves are made up largely of cellulose and so are hard to digest, and their cells contain little sugar. Trees understandably also try to protect these productive organs from herbivores. Once their leaves have expanded to their full size, they toughen them by adding extra cellulose and lignin to the ribs, which makes them harder to chew and protects their cell contents. Herbivores generally respond to this defensive strategy by eating only the young, expanding leaves at the tips of the branches. However, plants can retaliate by filling their young leaves with poisons, most commonly tannins and phenolics, which taste bitter and precipitate digestive enzymes in the guts of their consumers. So a leaf-eating primate has to eat huge quantities of young leaves and hold them for days in its stomach to detoxify and digest them; this limits its energy intake. Leaf-eating monkeys tend to be large, potbellied animals and have a slow metabolism and limited intelligence—they cannot afford to develop a large brain, but then again, as leaves aren't hard to find, they don't need to! The archetypal leaf-eating primate is the proboscis monkey of Borneo. These bizarre animals travel around in small groups, led by a dominant male whose weird looks give the creatures their name. They have long pink noses, markings like underpants around their groin, and most important a distended stomach. All of these reminded the local Indonesian people of Western colonialists, hence the common name *orang Belanda* or "Dutchman."

Those primates that changed their diet to eat fruit rather than leaves also tended to get bigger because fruit is plentiful in rain forests and is full of energy; however, a diet of fruit has also led to rather more profound changes in their brains. As a food, fruit has many advantages. Plants produce fruits as a reward—a way of persuading animals to ingest their seeds and eventually disperse them in their feces—so they fill them with sugars and soften them up as they ripen to make them easier to chew and digest. They even signal to animals that the fruit is ripe by changing its color and by developing an attractive odor for it. The only downside of eating fruit is that with so many different types of trees in tropical rain forests, each species is widely scattered through the forest. Moreover, because of the lack of seasonality, trees can fruit at any time. Trees that are in fruit are rare and hard to find. Fruit-eating primates not only have to be able to spot when fruits are ripe, but also have to be able to remember where fruiting trees are located within the forest, and to predict when they are likely to fruit, so they can get to them before the fruit is eaten by other animals.

Consequently fruit-eating animals have to hold a great deal of information in their heads, mapping the world in space and time. Field studies and experiments on captive fruit-eating primates have shown that they can remember the location of large numbers of fruiting trees and compute accurate routes to travel rapidly and economically to the next tree to ripen. So it is no surprise to find that fruit-eating primates such as macaques and spider monkeys have brains that are on average about 25 percent bigger than those of their leaf-eating cousins, the langurs and howler monkeys. This has enabled them to develop more sophisticated social behavior and live in more cohesive groups. Some monkeys, such as the capuchins, have even learned to make and use simple tools; they use stones as hammers to crack open nuts and shellfish.

But the intelligence of monkeys pales in comparison with that of our closest relatives, the great apes: orangutans, gorillas, chimpanzees, and bonobos, whose brains are twice as large relative to their body weight. Most primatologists believe the apes acquired their larger brains to help them communicate with and manipulate their peers. And they certainly do exhibit complex social interactions within their group; they seem capable of feeling empathy, have a self-image, and exhibit a degree of consciousness as they can recognize themselves in a mirror. But this "social hypothesis" does not explain why it was the great apes that became so clever, rather than monkeys or a group of terrestrial mammals. Nor does it explain why orangutans, who seldom encounter their neighbors, are so intelligent. It seems likely that some other factor must have been in play that caused apes to become more intelligent in the first place, and which could subsequently have enabled some members of the group to develop high-level sociality.

I first started to think about the evolution of ape intelligence many years ago when I was a young academic visiting the forests of Sabah, Borneo, for a quite different reason: to investigate why tropical rain forest trees develop huge platelike buttresses between their trunks and their roots. The research center where I was staying was also the base for some young British research students who were investigating what orangutans were doing with their huge brains. They were testing the hypothesis that the apes needed greater brainpower to map and predict when and where fruit would ripen across the forest.

I wasn't convinced. The hypothesis had already been used to explain why fruit-eating monkeys were more intelligent than leaf-eating ones, so it was unlikely to explain why orangutans that lived in the same forest as macaques, and ate much the same food, would have even larger brains. I approached the problem from my quite different perspective: as a biomechanic—someone who studies the engineering of plants and animals. And the most obvious physical difference between monkeys and great apes, apart from apes not having tails, was size; all the great apes are much bigger and heavier than monkeys. It is not immediately apparent what effect this would have on intelligence; after all, tigers are not more intelligent than wild cats; and capybaras, the world's largest rodents, are not more intelligent than mice. The difference, though, was where these primates lived: the forest canopy. A larger animal must have much greater difficulty moving around the canopy of trees, and in particular moving between trees, than a small one. The wooden branches would deflect more under their weight and would be more likely to break. The consequences of a fall for a larger animal would also be far more serious. As the great evolutionary biologist J. B. S. Haldane put it in his essay "On Being the Right Size":

"You can drop a mouse down a thousand-yard mine shaft; and, on arriving at the bottom, it gets a slight shock and walks away, provided that the ground is fairly soft. A rat is killed, a man is broken, a horse splashes."

An orangutan would probably be killed by a fall from the canopy that would scarcely harm a small monkey. It struck me then that the early apes might have evolved larger brains to help them navigate safely around their perilous arboreal environment and allow them to plan and follow the best routes through the trees. To do this they would also have had to develop a self-image; they would have to realize that their body weight altered their mechanical world by bending down the branches that were supporting them. In other words their intelligence had a physical basis, not a social one: a feeling for the mechanical properties of wood. I even wrote up my idea in a scientific paper, but was not surprised when it was rejected by a journal. After all, I was not a primatologist, I'd not been to the rain forest before, and I had no data on the actual behavior of apes. I was an outsider making an educated guess. I went back to looking at what I was meant to be studying, the anchorage of trees and other plants.

Many years later, though, I was surprised and pleased to learn that my idea was now a bona fide theory of the evolution of intelligence in apes—the "clambering hypothesis" of Daniel Povinelli and John Cant. These two American primatologists had also thought about orangutans, though in their case they had put in a good deal of time in the field observing these magnificent creatures. They had noticed the painstaking way in which orangutans travel between adjacent trees; they move slowly and gingerly, holding on to several branches at once with their hands and prehensile feet. Like me the two men reasoned that the apes could have developed a conception of self to allow themselves to move safely through the canopy. And since the publication of their hypothesis in 1995, other field-workers have built up evidence that orangutans, in particular, do have a high level of understanding of the mechanics of trees.

Susannah Thorpe, now at the University of Birmingham, has studied the locomotion of Sumatran orangutans over many years and has shown that they move quite differently when they travel along branches of different diameter. On a thick, rigid branch they walk on all fours on top of it or hang below the branch, swinging themselves along. In contrast, on branches less than 4 centimeters (1.6 inches) in diameter, they either clamber, holding their body horizontal and gripping several branches, or walk upright on their hind legs while holding on to branches above them with their hands. In both these cases, they distribute their weight between several branches, making their locomotion much safer. They can even exploit the flexibility of tree trunks by climbing high into the canopy and rhythmically shifting their body weight back and forth to make the tree sway so that they can reach across a gap to get to the neighboring tree.

Understanding the mechanics of tree branches gives the great apes another advantage: they can use them to construct a nest in which they can safely sleep. All the great apes are capable of making themselves complex cup-shaped nests in the tree canopy, though huge silverback male gorillas tend to stay on the forest floor. Such nest building provides unexpected benefits and opens up new opportunities.

Monkeys sleep on branches high in the forest canopy. This certainly keeps them safe from ground predators such as leopards and jaguars, but it must be precarious and uncomfortable. The monkeys sit on as thick a branch as they can find, resting their weight on pads of skin that develop on their buttocks, but even so they repeatedly wake up throughout the night. An ape, sleeping within a broad, cup-shaped nest, is far safer and can sleep for longer periods and more deeply. Studies by David Samson, now at the University of Toronto, and his coworkers, comparing neural activity in sleeping monkeys and sleeping apes, have shown that the apes have more frequent bouts of both NREM (non–rapid eye movement) and REM (rapid eye movement) sleep. These types of sleep are important in reordering and fixing memories, which can in turn help improve cognitive ability. Building nests could have helped apes get even cleverer.

It might seem to be a simple task to construct a nest, and that certainly appears to be what primatologists have thought, as they have given them scant attention. But it is not just a matter of breaking a few branches off and weaving them together. For a start, as any gardener knows, and as I was taught as a Cub Scout collecting wood for our campfire, it is nigh on impossible to snap a living branch off a tree by bending it. And this is not because the branches are too strong, but because the structure of wood affects how it breaks.

Wood is quite a complex material, but the main factor that affects the way it fails mechanically is the macroscopic arrangement of the wood cells. Most wood cells are oriented longitudinally up and down the trunk and branches: the long, narrow tracheids that give wood its strength; and the lines of broad open vessels that in broad-leaved trees conduct water. The only other cells are the ray cells, which form spindle-shaped rays that run radially from the pith to the bark, and which reinforce the trunk in this direction, effectively pinning the growth rings together, and preventing the trunk from falling apart.

This complex structure gives wood different mechanical properties in different directions. Wood is hard to break across the grain because this involves fracturing the tracheid walls, whereas it is easily split along the grain, as this just involves separating the tracheids from each other and breaking a few ray cells. Splitting a branch radially is particularly easy as the fracture runs between the rays. Wood is consequently eight to ten times stronger longitudinally than transversely, and most types of wood are also 20–50 percent stronger in the radial direction than in the tangential. This pattern matches the forces the wood has to withstand. The high strength and stiffness of wood along the grain enables it to withstand the bending forces to which tree trunks and branches are subjected by gravity and the wind.

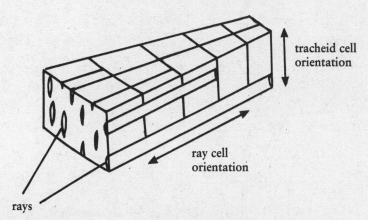

Structure of the trunk of a conifer. The tracheid cells run up and down the trunk, while the ray cells run radially from the center of the trunk to the bark through the growth rings.

The longitudinal fibers are ideally arranged to resist the longitudinal tension and compression forces that the bending sets up within the branch.

But this structural arrangement also makes it almost impossible to detach a living branch. If you bend a branch of green wood, what you are doing is stretching the wood on the convex side, and compressing the wood on the concave side. In a typical branch the wood will fail first in tension, and the branch will start to break across, like a carrot or stick of celery. But it won't break all the way. As the crack reaches the center of the branch, it gets diverted, traveling up and down the weak center line of the branch, between the tracheids and between the rays. Despite your best efforts, the branch will split along its length and remain half attached. A similar sort of failure occurs in the long bones of children, when it is known as greenstick fracture—coincidentally this often occurs when they fall out of trees! I tasked my PhD student Adam van Casteren, who was out in Sumatra studying how orangutans use the flexibility of

How branches fail when you bend them. They break halfway across, but then split along their length, a process known as greenstick fracture.

branches to move through the trees, to investigate how these creatures overcome the problem of greenstick fracture to build their nests.

Working in the rain forest of Aceh, Indonesia, Adam followed the apes during the day and observed them making their nests in the evening, returning the next morning to climb up to the nests, examine them, and perform mechanical tests on the structures. What Adam found—backed up by films of nest-building orangutans taken by Julia Myatt, Susannah Thorpe's PhD student—was that an orangutan would find a good strong horizontal branch to rest on, then construct its nest around this support. First, it would lean out and with one hand draw thick branches in toward itself, breaking them in greenstick fracture and hinging them inward, before finally weaving the branches together. The result was a cup-shaped elliptical nest around four feet long and two and a half feet wide. Sitting in the completed structure, the ape would reach out to grab thinner branches and, holding them in two hands, first break them in greenstick fracture, then twist them to break the two ends apart. It then stuffed the broken branches, complete with twigs and leaves, into the nest, behind and around itself to produce a mattress and a pillow, and finally on its lap to produce a blanket. The whole process was remarkably rapid. In Julia's film, the male ape took only five minutes to build his nest, and half of that time was spent resting between the two stages. Clearly, by the time they are adults—and it takes young orangutans years of observing their mothers and practicing by themselves for them to perfect their constructions—orangutans have an excellent working knowledge and "feel" for the mechanics of green wood.

Given the sophisticated nest-building behavior of apes, it

is no surprise to find that they are also capable of making and using simple wooden tools, though primatologists have been reluctant to link these two capabilities. This reluctance may be due to their overreliance on strict definitions. Primatologists define tools as items used for a particular function that are detached from their environment and usually held in the hand. Nests are clearly not tools in these terms, though they require at least equal skill to make. Whatever the reason, it is unfortunate, as it has meant that until recently primatologists have failed to realize the importance of nest-building behavior in the evolution of toolmaking.

Compared with other apes, orangutans make rather few tools in the wild. They have often been observed breaking off twigs and using the ends to poke into holes to collect termites, and Carel van Schaik of the University of Zurich, Switzerland, found a dense population in the swamp forests of Suaq, Sumatra, that had developed two different sorts of wooden tool: one to extract honey from within hollow trees; and a second to prize open the shells of *cemengang* fruit to extract the nutritious seeds. Van Schaik even found that the orangutans changed their tool design through the season, choosing wider sticks later on as the fruit's cases gradually opened up. Other populations of orangutans are not so innovative, probably because of a lack of incentive and opportunity: they have little need of tools to eat their normal diet of fruit, and being more solitary, there are few opportunities for a toolmaking culture to develop. In captivity, in contrast, orangutans are notorious for their manipulative skills: they can take scientific apparatus apart with ease and escape from the best-designed cages.

In the wild, chimpanzees are the most proficient and inventive users of wooden tools. Many groups of chimps make fish-ing sticks like those of orangutans to extract termites, and a chimp may use two different sticks for the purpose: a strong thick stick to make the hole, and a thinner stick with a frayed end to extract the termites from the hole. The honey-loving chimps of Gabon are even more sophisticated. Christophe Boesch of the Max Planck Institute for Evolutionary Anthropology, Leipzig, found that they have developed and carry around a whole wooden-tool kit to break into bees' nests and raid them of honey. There are thin perforators to probe for the nests; blunt, heavy pounders to break inside; leverlike enlargers to widen the holes and access the different chambers; collectors with frayed ends to dip into the honey; and swabbers (elongated strips of bark) to scoop it out.

But it is in the most extreme environments where chimps make the most innovative tools, and the ones that remind us most strongly of the tools used by modern hunter-gatherers. The savanna chimps of Tanzania, East Africa, make and use digging sticks one to two feet long to probe into the soil during the wet season to excavate plant tubers. The savanna chimps of Senegal have even more disturbingly humanlike powers. Jill Pruetz of Texas State University has observed female chimps making and using spears. They break off two-to-four-foot-long branches, strip them of leaves, and sharpen the narrow end with their teeth. They use these tools to probe into hollow tree trunks, flushing out and even spearing bush babies, which they then eat to supplement their vegetarian diet.

The great apes have clearly made great advances in their mental abilities since they split from the monkeys. This has enabled them to cope with the flexibility and weakness of the branches among which they live, to build complex wooden nests, and to make wooden tools that are in many ways more sophisticated than the stone tools used by early humans. Our earliest ancestors, who split away from the line leading to the chimpanzees and bonobos some 5–7 million years ago, most likely shared these abilities; they would have been architects and artisans whose chosen material was wood.

But despite our similarity to the great apes, there is one ability we still regard as unique to ourselves: the ability to walk upright on two legs. Television films of bonobos walking waist-deep in water look uncanny and disturbing. Most apes can only walk bipedally for short distances, and then in a crouched posture with bent legs and a forward-leaning stoop. Chimps and bonobos tend to move across open ground on all fours, touching the ground with the knuckles of their hands, rather than with the palm, so-called knuckle walking. The one great ape that does walk more or less upright and with straight legs, just like ourselves, is surprisingly the most arboreal one, the orangutan.

And evidence is starting to build that the evolution of bipedalism did not take place on the ground or in the sequence depicted in the conventional picture, through the intermediate stage of knuckle walking. Instead, there is support for an alternative hypothesis put forward by Susannah Thorpe and by Robin Crompton of the University of Liverpool, that our ancestors gained their ability to walk bipedally when they still lived in the trees. Moreover, it is becoming clear that far from striding out immediately into the plains, our ancestors remained in well-wooded regions and stayed in the canopy long after they had become able to walk upright.

Much of the evidence for this hypothesis comes from studies on living apes, and in particular Susannah Thorpe's investigations of the way orangutans move in the canopy. We have already seen that orangutans frequently walk upright along narrow branches, and that when they do so, they also cling to higher branches with their hands. This allows the animals to distribute their weight over more than one branch and so to travel more safely. But it could also allow the orangutans to take advantage of the springlike flexibility of the branches. As an animal puts its foot down, the branch moves downward under its weight, storing energy, before springing up again and returning that energy. The orangutan could therefore bounce along the branch almost effortlessly, like a person walking on a trampoline. Adam van Casteren tested this aspect of the arboreal bipedalism hypothesis by investigating the mechanical behavior of branches, and seeing whether orangutans could and did use them like springs to help them walk more efficiently. He measured the stiffness of large numbers of branches along their length and investigated how quickly they could spring back up again if an orangutan was standing on them. The behavior of the branches was complex, but Adam found that the thickness of a branch at any particular point predicted its stiffness well; an orangutan would be able to tell how stiff a branch was just by looking at how thick it was. The branches also swung back and forth quickly enough to return energy to walking orangutans. Adam even filmed a few occasions when orangutans were bouncing their way along a branch, something that Susannah was able to replicate more easily with captive orangutans at Chester Zoo, England, filming them walking along purpose-built beams.

Susannah and her research assistant Sam Coward also showed that holding on to branches could help an animal overcome another major difficulty of evolving bipedalism: keeping its balance. This time they used humans as a model species.

Experimental subjects had to balance on a springboard while an image of trees swaying slightly in the wind was projected around them. Half the time the people had a flexible pole to hang on to, mimicking a handhold on a branch, while at other times they had no handhold. The people were filmed, and the neuronal activity in their thigh muscles was measured when the springboard was perturbed. The researchers found that following the perturbation people's thigh muscles had to work harder to maintain balance, but having the handhold reduced this work by up to a third; it had helped the people balance more efficiently.

So being able to walk upright in trees had clear benefits, and the fossil evidence shows that our ancestors did show a gradual change in their lower limbs that allowed them to do so, even while the rest of their bodies were still adapted to arboreal life. For instance, *Orrorin tugenensis*, one of the earliest hominins on the fossil tree, which lived around 6 million years ago, had the head of its femur bent inward like that of a modern human, suggesting it was capable of bipedality. However, it still had fingers and toes that were curved inward, adaptations to holding on to the branches of trees. The hip and leg bones of the 4.4-million-year-old *Ardipithecus ramidus* were even better adapted to walking upright, but its feet still had opposable big toes, like a great ape, so it was also well adapted to quadrupedal walking and clambering in the forest canopy. And a recent find in Germany shows that this ability to walk upright in the tree canopy might have evolved much earlier. The 12-million-year-old fossil ape *Danuvius guggenmosi* had lower limbs like those of *Ardipithecus*, suggesting that bipedalism might repeatedly have evolved among the great apes.

Paradoxically, therefore, our ancestors developed the attributes, both physical and mental, they would need to succeed on the ground while they were still in the forest canopy. As we shall see, though, the move to the open was still not complete. The next chapter shows how we were able to use our relationship with wood to finally come down from the trees, keep our feet on the ground, and become truly human.

CHAPTER 2

Coming Down
from the Trees

In 2016 the anthropological community was shocked to learn that one of its most famous daughters, Lucy, had met a violent death falling from a tall tree. Lucy was not herself an anthropologist, but the most celebrated of all early human fossils. She was a member of the early hominin species *Australopithecus afarensis,* and much of her skeleton had been found in 1974 among 3.2-million-year-old rocks in Ethiopia, by Donald Johnson of the Cleveland Museum of Natural History. Named Lucy, after the Beatles song "Lucy in the Sky with Diamonds," which was playing at the anthropologists' camp, she quickly became a star, as it became apparent that she was capable of walking upright just like us. She had humanlike hip bones with short iliac blades and a wide sacrum, and the top of her femur bent inward toward the hip joint, allowing her legs to point down vertically. Further biomechanical studies of Lucy over recent years have confirmed the initial interpretation that she could walk like a modern human. Bill Sellers of the University of Manchester reconstructed Lucy's lower body in a computer and produced simulations of a walking gait that were essentially human. And in 2011 Robin Crompton of the University of Liverpool found that the footprints left in sand by even-earlier australopiths some 3.6 million years ago strongly resembled modern footprints. Clear impressions of the heel and the ball of the foot showed that these creatures had adopted our characteristic straight-legged walking gait. Combining such overwhelming evidence with the location of the fossil finds, the drought-ridden Afar Triangle of Ethiopia, it is not surprising that the earliest reconstructions of Lucy showed her striding across a barren landscape of grasses, with just the occasional bush in sight. Lucy, it seemed, was the first girl on the catwalk of human evolution; so the idea of her being in a tree, let alone falling from it, might seem unlikely.

But the evidence for the scenario of Lucy's death is fairly compelling. It had always been assumed that the breaks in her fossilized bones had occurred in the millions of years since she died. However, when John Kappelman and his coworkers from the University of Texas scanned her skeleton in an MRI machine, they found fractures characteristic of the sort that are seen in adult fall victims. The leg and arm bones both had complex compression fractures, with breaks running at forty-five degrees to the long axis of the bones. They also found instances of the greenstick fractures that are seen in children who have fallen out of trees; the bones were fractured in bending, but had only broken halfway across before splitting lengthwise, just as when green twigs are bent. And Kappelman's interpretation that Lucy had fallen from a tree fits in with a number of other discoveries that have suggested that far from being fully terrestrial, Lucy and her relatives were semiarboreal. For a start, the environment of East Africa was not then as arid as it has since become; in Lucy's day the area would have been covered with savanna woodland.

There is also compelling anatomical evidence from the upper

bodies of australopiths that Lucy was semiarboreal. She had strong chimplike arms and curved fingers that would have been ideal for a life involving a lot of tree climbing. In 2012 David Green of Midwestern University and Zeresenay Alemseged at the California Academy of Sciences also showed that she had apelike shoulder blades. And in 2016, CT scans of Lucy's bones by Christopher Ruff of Johns Hopkins University and colleagues showed that she had thick-walled arm bones like those of chimpanzees and unlike the thin-walled bones in our own arms. She must have used them for climbing. Finally, a 2018 study of the foot bones of a juvenile *Australopithecus afarensis* from 3.3 million years ago showed that this youngster had even more curved metatarsal joints than Lucy, something that would have enabled it to move its big toe in and out, so it could have used it like a thumb to grip branches. It must have spent plenty of time climbing in the trees and clinging to its mother.

So, though early australopiths such as Lucy looked like us from the hips down, above the waist they would have resembled apes. The same seems to have been true for even more recent hominin species. *Australopithecus africanus*, which survived until just over 2 million years ago, also had the long arms and curved fingers of a tree climber. And even the first member of our own genus, *Homo habilis*, which lived between 2.1 and 1.5 million years ago, had relatively longer and stronger arms than us. It seems that it was only with the emergence of *Homo erectus*, less than 2 million years ago, that humans became fully adapted to a terrestrial lifestyle.

But if hominins had been able to walk upright like us 3.6 million years ago, why did they retain their ability to climb trees for a further 2 million years? To understand the human story we have to be able to explain not only why early hominins came down from the trees in the first place, but why they were so reluctant to make the descent permanent. And we also need to be able to explain how *Homo erectus* finally managed to break free to become truly terrestrial.

Recent studies of the history of the global environment have shown that the key to understanding why hominins came down from the trees is climate change. Over the last 20 million years, the world's climate has been getting cooler, caused in large part by movement of the earth's tectonic plates. As India has plowed into the Eurasian plate, it has forced the Himalayas upward, and the silicate rocks this exposed have absorbed carbon dioxide from the atmosphere, reducing the strength of the greenhouse effect. The climate has cooled, and the tropics and subtropics have become more seasonal, with wet seasons being interspersed with increasingly long dry seasons. This trend has been most evident in East Africa, where the mountains that have been pushed up by the formation of the Great Rift Valley have cut off the rainfall coming from the Indian Ocean. What previously was continuous tropical and monsoon forest has opened up, the trees unable to cope with the longer dry seasons except in the damper soils along river valleys.

The exposure of the forest floor to light has allowed a new type of ecosystem, savanna, which is dominated by grasses and other herbaceous plants, to take over. These plants only grow in the wet season and survive the dry season by dying back and storing their energy underground in fleshy bulbs, corms, or roots.

Clearly, this change in vegetation was bad news for forest-dwelling apes. Like the modern-day savanna chimps that we saw in the last chapter, they would have been forced to the forest floor, first of all to travel between the scattered trees, but also in search of other types of food to supplement their diet of fruit.

They must have varied their diet, as modern-day chimps do, by eating the termites that abound in savannas, raiding honey from bees' nests, and hunting small mammals such as bush babies. Like the chimps, they probably fashioned wooden tools, such as probes, chisels, and spears to do this, and maybe used stone hammers to break open the hard nuts and seeds that the new types of drought-tolerant plants produced. But their main source of food in the dry season, like modern-day hunter-gatherers such as the Hadza people of Tanzania, who live in similar savanna woodland, would have been underground roots and bulbs.

Unfortunately, eating roots presents difficulties because plants receive no benefit from their roots being eaten; unearthing a plant's roots and eating them kills the plant! In contrast to fruits, therefore, which are adapted to be easy to eat, roots are strongly defended. First, plants protect them mechanically, by incorporating tough fibers within them. Both the early australopiths and *Homo habilis* developed their dentition to cope with these mechanical defenses; they replaced the pointed canines and cusped molars of fruit-eating apes with smaller canines and huge platelike molars coated with thick enamel: teeth more suited to breaking off and grinding up tough plant material. Later australopiths, such as *Paranthropus boisei* and *Paranthropus robustus*, also developed large sagittal crests on the top of their heads, rather like ones you can see on modern hyenas, which acted as the insertion points of huge jaw muscles. It is thought that this would have helped them grind up the tough roots and crack open hard nuts and seeds.

Plants also defend their underground storage organs chemically, by incorporating astringent chemicals to precipitate out digestive enzymes, and toxins to poison consumers. Australopiths developed large guts to help digest this difficult food, something shown by the expansion of their lower rib cage. They must have been potbellied, just like proboscis monkeys. But the main difficulty in eating roots is accessing this subsoil resource in the first place. Baboons, the only primates that currently live on the African plains, use their hands to dig in the soil, but they can only reach shallow bulbs and corms. Warthogs use their impressive tusks to dig a bit deeper. The hominins would have had to develop a new technology to access even longer, deeper roots. As we saw in the last chapter, some modern savanna chimps use digging sticks to do this, but these are seldom more than 0.6 inches thick and 12 inches long. With such short, spindly tools, which for some reason they hold at the thin, weak end, they are incapable of digging up anything but shallow roots and bulbs, and then only in the wet season when the soil is soft. Australopiths would have needed something better.

No experimental studies have been carried out to investigate the best design for digging sticks; fortunately, though, the prizing out of soil is mechanically similar to the felling of taprooted plants, something that I myself have investigated. Simple mechanics tells us that to improve their digging performance, early hominins would have needed to break off and use longer, stouter sticks; a stick twice as thick is sixteen times as rigid and eight times as strong; when used for digging, such a stick would be able to prize out double the depth of soil. The hominins would also have had to sharpen the end of the stick to allow it to be pushed into the soil more easily. Bearing in mind the perishability of wood, it is not surprising that no digging sticks have been found from the period of early hominid evolution. Indeed the earliest digging sticks that have been found are only 170,000 years old and were made by our close relatives the Neanderthals at Poggetti Vecchi in southern Tuscany.

They were forty to fifty inches in length, one inch to one and a half inches in diameter, and were sharpened by charring the narrow end in a fire and scraping off the burned wood. The digging sticks used by modern-day hunter-gatherers, such as the women of the Hadza tribe, are even larger and more sophisticated. They cut sticks that are over a yard in length, an inch and a half thick, and weigh anything from one to two pounds. And the roots the Hadza dig up using these tools are extremely impressive, putting our carrots and parsnips in the shade. Their favorite //ekwa hasa roots are around four feet long and highly nutritious. The Hadza women dig them up by pounding the pointed end of their sticks into the soil to break it up and levering out the loosened soil with a digging motion; the process is so efficient that the women can collect enough roots in a few hours for the daily needs of their band.

These sticks are quite sophisticated implements, but the Hadza have access to iron tools such as machetes, which they use to cut new sticks every week or so and to sharpen their tips. As they did not have access to metal or, in the earliest times, perhaps even stone tools, australopiths would have been unlikely to be able to make digging sticks as large as those of the Neanderthals or the Hadza. However, there must have been strong selection pressure in early hominins to learn how to break off and prepare thicker, longer, and stronger sticks. This may have driven them to develop new stone tools with sharp edges that could saw through wooden branches and whittle the ends into points. To do this, and to handle the digging sticks effectively, they would also have had to evolve stronger gripping hands with fully opposable thumbs.

In using digging sticks, early hominids would have made use of the superior mechanical properties of wood. We saw in the last chapter how the cellular structure of wood influences how a branch breaks when it is bent. But its strength, stiffness, and toughness is down to the molecular structure of the cell walls themselves. The cell walls are stiffened by crystalline microfibrils of cellulose, which are embedded in a softer matrix of hemicellulose that is stabilized by a polymer called lignin. The beauty of the design is the way that most of the fibrils coil around the cell at an angle of around twenty degrees to its long axis, reinforcing the cell along its length. And when the cell wall finally breaks, the fibrils uncoil like a stretched spring, creating a rough fracture surface with thousands of tiny hairlike fibrils projecting out of the wood. This process absorbs huge amounts of energy, making wood around a hundred times as tough as fiberglass, and giving wood its resistance to fracture. It's the reason why trees stand up so well to hurricanes that can destroy more rigid man-made structures, and why wooden boats are far more resistant to bumps than fiberglass ones.

But the early hominins would also have been helped by the first of two incredibly fortuitous properties of wood, properties

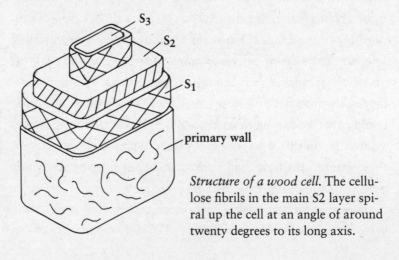

Structure of a wood cell. The cellulose fibrils in the main S2 layer spiral up the cell at an angle of around twenty degrees to its long axis.

that are of no actual benefit to the trees that make it. If wood is broken off a tree and starts to dry out, its mechanical properties improve! This is most unusual for biological materials; bones, horn, and nails all get weaker and more brittle as they desiccate. As the cell walls of wood dry out, however, and water evaporates from within the hemicellulose matrix, the matrix stiffens up, preventing the cellulose fibrils from shearing past each other so easily. This stiffens the wood while leavings its strength and toughness, which depend on the cellulose fibrils themselves, unchanged. At the 60 percent relative humidity of the savanna dry season, the water content of wood typically drops from 30 percent to 12 percent and its stiffness triples. Early hominins would have made use of this transformation, just as people have been doing ever since. They could have sharpened the points of their digging sticks while they were still green, using either their teeth or sharp stones, and used them later when they had dried out and stiffened up. A fully dried stick would be able to dig a hole around 50 percent deeper than a green stick.

It is perhaps best, therefore, to visualize early hominins as essentially bipedal and semiarboreal apes. Their brains were not much larger than those of modern-day chimps; Lucy's brain had a mass of around one pound, the later *P. boisei* and *P. robustus* had brains of around 1.2 pounds, and *Homo habilis* had a 1.4-pound brain. They must have behaved rather like modern-day savanna chimpanzees, eating a wide variety of vegetable matter, but with rather greater reliance on underground roots, and a greater ability to fashion and use wooden tools. Certainly, by 3.2 million years ago they were also using primitive stone chopping tools, the so-called Oldowan tools, which we will look at in more detail in chapter 4, and they had also evolved hands that were better for grasping these tools. However, they were

, almost certainly still covered in hair and had upper bodies that would enable them to climb back up into the forest canopy, with strong arms and shoulders and curved, grasping fingers. Lucy and her relatives must have lived in savanna regions, with the grassland interspersed by areas of forest, or single trees. But if they were foraging so much on and beneath the ground, it seems puzzling that they continued to return to the trees; there must have been a major problem that prevented them from coming down permanently.

Looking at the present-day African plains, it is clear what that problem must have been: they would have been extremely vulnerable to being eaten by predators such as saber-toothed cats, scimitar-toothed cats, and the ancestors of present-day lions and hyenas. Today, baboons are the only large primates that live on the plains of Africa, and they have real problems with predation. Compared to early hominins, they are physically far better able to defend themselves; they have huge canine teeth, and a fully grown male may weigh as much as ninety pounds, more than a match for many large cats. Even so, baboons have to live together in groups of twenty to two hundred individuals to protect one another, and yet they still get a rotten night's rest. The sleep specialist David Samson and his colleagues have found that even when they are living in zoos, baboons wake up eighteen times a night, only sleep for 60 percent of their rest period, and get into deep REM sleep only around 10 percent of the time. This contrasts with 18 percent of the time for chimpanzees, which sleep in nests, and 22 percent for modern humans. Considering the importance of deep sleep for the functioning of a large brain, this would mean that for a relatively defenseless early hominin, being able to climb trees would have been vital to enable them to survive and evolve intelligent behavior. Like savanna chimps they must have carried on building nests and sleeping in the canopy. They would have climbed into the canopy not only in search of fruit, to break or saw off digging sticks, but also to make sleeping nests, where they could rest safely overnight. Lucy's death was probably just an accident as she was going about her morning or evening chores.

So how did early humans finally come down permanently from the trees? The only plausible way that our ancestors could have protected themselves on the ground at night from predators was by using fire. This is where the second of wood's fortuitous properties comes in: it is flammable, especially when dry, and when it is burned, it releases a large amount of heat and light. The flammability of wood is of no use to trees; it's just another fortunate accident that it does burn, though most living trees, especially ones growing in rain forests, are extremely resistant to being set alight. As we saw earlier, the cell walls of living wood contain a lot of water, around 30 percent of their dry weight. Moreover the cell lumens in the sapwood around the outside of the trunk and branches are filled with water; a tree trunk can therefore contain three times its dry weight of loose water. Before wood can burn, all this water has to be heated up and evaporated off, which requires as much as a third of the energy that is released when the wood finally burns. If you put a branch of green wood onto a fire, it takes some time to heat up before it eventually starts spitting and hissing, releasing water and steam from its ends. The evaporation of water within the cells is what destroys trees when they are hit by lightning. The electrical energy from the strike heats up the water, making steam that expands explosively and splits the trunk apart, long before the steam can escape through the tips of the branches. Lightning seldom ignites fires in trees directly, though; if it starts forest fires, it does so by setting light to the dry grass and twigs around the trees.

The processes involved when dry wood burns are quite complex and bear a detailed explanation, since they determine how we can build and control fires. Cellwall material is chemically stable, even at temperatures above 212°F; the lignin keeps the cellulose fibers rigidly bound together, which explains why we can't cook wood and make it into a useful food by boiling it! Above the boiling point, the bound water evaporates from wood cells, but no other change occurs until the temperature reaches 300°F. Then hemicellulose molecules between the cellulose fibers crystallize, making them stiffer. This also hardens the wood itself, but since it prevents the cellulose fibers from pulling out of the cell wall, it also makes the wood much more brittle. We shall see in chapter 4 what use people have made of this "fire-hardening" effect.

Only at around 400°F does heat start to break the wood down. The huge polymer molecules—cellulose, hemicellulose, and lignin—start to split up and to form a wide range of smaller liquid molecules. This process, known to scientists as pyrolysis, releases energy, which for the first time starts to generate heat to drive the burning. As the temperature rises further from 400°F to 600°F, these small molecules evaporate, and some of them react with the oxygen in the air to produce a flame, generating further heat. Some of the gases escape, however, along with some carbon particles, and are released as smoke. Finally, when the breakdown of the cell wall has been completed, only carbon is left; the wood has been transformed into charcoal. Unlike the volatile chemicals produced by pyrolysis, the carbon does not evaporate and only burns when the temperature reaches 900°F; it reacts with oxygen at its surface to produce carbon dioxide and energy. Since nothing evaporates from the charcoal, however, no flame is produced and there is no smoke, which is why the embers of a fire just glow red-hot.

Now that we understand combustion, we can see that the key to starting a wood fire is to raise the temperature of the wood high enough to allow pyrolysis to start, while allowing oxygen to reach the surface to sustain the fire. We also need to place enough wood material nearby to allow the fire to spread. This is why the first step in laying a fire is to make a loose pile of tiny pieces of kindling, which can readily be heated up and catch fire. The fire can then be stoked with larger and larger pieces of wood, which will heat up, undergo pyrolysis, and burn in turn, before the fire is hot enough to make logs glow red-hot, at which point the fire becomes self-sustaining.

Starting fires without matches or firelighters is no easy business. The usual methods employed by modern hunter-gatherers are either to generate heat by rubbing sticks together, or to make sparks by striking flints against each other. It is unlikely that early hominins would have been able to do either. Fortunately, they would have been helped by several factors. The first is the climate change that was driving them down to the ground in the first place. In the dry season of savannas, the water content of the air falls to only about 60 percent of the saturated value, and the hot sun and wind rapidly dry deadwood until it has only around 12 percent bound water in its cell walls; it becomes far more flammable. Savannas are also prone to frequent naturally occurring bush fires, which are set off by lightning strikes and kindled by dead grass. Today, predators such as cheetahs and birds of prey are drawn to bush fires, feeding on the small mammals and birds that are flushed out in a panic by the flames. Savanna chimpanzees are also attracted to fires, gathering the cooked seeds of bean trees and eating them. Early hominins would have been able to mirror this behavior and, like modern hunter-gatherers, could have collected and eaten a wider range of tree seeds and nuts that had been opened and partially cooked by the fire.

From following and using naturally occurring fires it is a small step to keeping those fires alight. Australian aborigines have a long history of using fire to manage their environment, setting light to areas of bush to cook seeds and insects and provoking new growth of edible plants. By so doing they have transformed their landscape, promoting the growth of the fire-resistant *Eucalyptus* trees that now dominate the Australian bush. To do this, they simply carry smoldering logs with them as they travel about the bush, lighting fires when they need them. And from keeping fire alight in smoldering logs, it is only another small step to keep a fire burning at a permanent camp, and building it up at night to repel predators. For early hominins, it would at last have been safe to stay permanently on the ground.

Setting up a permanent camp, and being able to sit together around the campfire, would have had other advantages. It would help to keep the hominins warmer during the cool nights typical of savanna regions. The light from the fire would also help lengthen the time when individuals could carry out tasks such as making and mending tools. There would also be opportunities for a greater variety of social interactions: sharing food and exchanging information. Having a permanent fire would help speed up the evolution of both practical and social skills.

But perhaps an even more important benefit of a campfire was that it could be used to cook food. In his 2009 book, *Catching Fire*, Richard Wrangham has put forward a convincing case that cooking would have been a crucial step in the evolution of modern humans—one of the main stages that took us from being semiarboreal, bipedal apes to being more or less human. As he points out, cooking meat and vegetables has two main effects—on their mechanical properties, and on their chemistry—that would have enabled hominins to radically change their digestive apparatus and behavior.

Cooking food heats it to well over the temperature at which it works in living organisms and breaks down its structural components. In meat, the most important structural material is collagen, a long-chain protein, whose ropelike molecules are joined together into sheets that provide a framework for the muscle cells they surround. These sheets make up the white marbling you can see in a raw joint of meat, and in life they transmit the forces the muscle cells produce to the tendons at the end of the muscle. Heating meat breaks up the collagen molecules and therefore weakens the sheets. This tenderizes the meat, especially the tougher (and consequently nowadays cheaper) cuts, which are made up of smaller lobes of muscle fibers and have more marbling between them; stewing steak, for instance. The more expensive cuts of meat, fillet and rump steaks, need less cooking because they have larger muscle lobes and contain less collagen. Cooking also softens plant material; heat breaks down the pectin that sticks the cell walls together and weakens the hemicelluloses that hold the cellulose fibers in the cell walls. However, heat alone cannot break down the cellulose fibers or remove lignin from within the cell walls of woody cells. This is why you can't soften the stringy fibers in grasses or overripe beans however long you cook them.

The consequences of the mechanical breakdown of food by cooking are profound. Both the stiffness and toughness of the food is drastically reduced, making mechanical processing far easier. The forces your teeth need to apply to break up the food are lower, and the food releases its cell contents far more easily. The optimal tooth shape for breaking up the food also changes. Rather than needing thick, flat plates that can apply large compression forces to grind up tough materials and crush hard ones, it is better to have molars with pronounced cusps to cut through the softer food. It yields at much lower forces and breaks up far more quickly. Modern hunter-gatherers spend far less time chewing their food even than apes that eat relatively soft fruit; they chew for less than an hour a day compared with five or six hours for chimpanzees. This frees up plenty of time for other tasks, such as looking after the fire, making a permanent camp, toolmaking, or further foraging.

But the chemical breakdown of food that occurs when it is cooked is even more important. Tests on modern humans have shown that we absorb far more of the energy within food when we cook it than when it is raw: something like 80 percent versus 60 percent. Furthermore, we need around 12 percent less energy to digest the food, while the time we take to digest it is halved. An early hominin that could cook its food would be better able to survive, reproduce earlier, and build up a higher population size. With faster, more efficient digestion, it would not need so large a gut and could divert the energy to produce and maintain a larger brain. Just as fruit-eating monkeys can have larger brains than leaf-eating ones, so hominins that cooked their food would have been able to support bigger brains than their relations that continued to eat raw food. The advantages of cooking are perhaps best shown by what happens to those human health fanatics who eat only raw food. Even if they grind up their food carefully before eating it, raw foodists have problems in digesting what they eat and invariably lose weight and conditioning.

Typical weight loss is around forty-four pounds for men and fifty-five pounds for women; and over half of fertile women stop menstruating, a clear sign of poor health.

The advantages of using fire are clear, therefore, but what is less clear is exactly when our ancestors started to make fires to protect themselves and to cook their food. The fossil evidence, in the form of changes in the teeth and skeleton that would be expected in a hominin that cooked its food, points to an early date of around 2 million years ago. This was when the first hominin that looked recognizably human, *Homo erectus*, appeared. As a candidate for the first user of fire it ticks all the boxes. For a start it had evolved changes in its skeleton that showed that it stayed permanently on the ground. It had an upper body similar to that of modern humans and, with its weaker shoulders and straight fingers, was plainly poor at climbing trees. It must have had ways of protecting itself at night when it was on the ground. That this method was using fire, and that it also used fires for cooking, is consistent with the changes that occurred in the skull and particularly its teeth. In *Homo erectus* the flat, platelike molars of australopiths and *Homo habilis* had been converted into smaller, cusped teeth remarkably like our own, and only capable of dealing with soft cooked food. The jaw muscles were also greatly reduced. *Homo erectus* no longer had the sagittal crest and large muscle insertions of the australopiths. Finally, the shape of its hips was much like our own, suggesting that it no longer needed to support a potbelly to help digest raw food. Most striking, though, *Homo erectus* also had a much larger brain—its brain mass ranged from around 1.9 pounds in early forms to 2.6 pounds in the most recent fossils—far larger than the other hominin species with which it would at first have coexisted. Feeding such a brain would only have been possible had it eaten easily chewed and digested cooked food, and it would also have enabled it to greatly expand its social and technical skills.

The fossil evidence to date the first use of fire at around 2 million years ago is strong, but there is less direct evidence in the form of the remains of fires that are associated with early humans. This is not surprising. The traces of most fires disappear in days or weeks, while caves, where the remains of fires might be expected to last longer, rarely last more than a quarter of a million years; they get weathered out of the bedrock. Nevertheless, a surprising number of sites contain evidence of burned wood that is associated with humans from 1.5 million years ago. These sites are in East Africa at Koobi Fora and at Chesowanja, while the first totally conclusive evidence of human-controlled fires was found at Gesher Benot Ya'aqov in Israel. At this seven-hundred-thousand-year-old site, charcoal and wood were found at a number of levels, together with burned flints and pebbles, marking out potential hearths.

The jury is still out, therefore, about when humans first used fire. But whatever the exact timing, what comes out most strikingly from modern research on early humans is that the key to becoming terrestrial was to make use of wood, and in particular to exploit two of its fortuitously useful properties. In the first stage of becoming terrestrial, early hominins would have made use of wood's becoming stiffer as it dries out, to fashion and use digging sticks to obtain their new source of food: underground storage organs. In the second stage, early members of our own genus *Homo* would have made use of the flammability of dry wood to make fires that could protect them from predators and cook their food. Our escape from the trees was paradoxically aided by a burgeoning relationship with the material they make: wood.

CHAPTER 3

Losing Our Hair

The Naked Ape was a brilliant choice of title for Desmond Morris's popular 1960s take on human behavior. Not only was it titillating enough to generate huge book sales, but it also perfectly captured what we feel about ourselves: if one physical feature above all others separates us from the apes, it is the lack of hair on our bodies. Indeed, hairlessness seems to have such a hold on our imaginations that modern society demonizes body hair. Sequences of hair growing on the backs of our hands are a staple of werewolf movies and are a mark of the bestiality of Robert Louis Stevenson's Mr. Hyde, while the global hair-removal industry was worth an estimated $880 million in 2017. But hairlessness is an extremely unusual trait for terrestrial mammals: only the naked mole rat comes to mind. So narratives about the evolution of modern humans really do have to explain when, and especially why, we lost our body hair. As we shall see, the answers are by no means fully clear, but once again it is likely that our relationship with wood must have played a major role.

Because hair does not fossilize, it is impossible to know exactly when our ancestors lost their body pelt. However, modern molecular genetics has been able to cast some light on the subject by looking at a gene that controls not hair growth, but skin color. One of the consequences of losing our body hair (or to be more exact, reducing the length and thickness of our hair follicles) was to allow more harmful sunlight to reach our skin. Since apes have rather pale skin beneath their hairy coats, a newly hairless hominin would have had to produce more melanin to absorb the harmful ultraviolet rays, turning its skin black. This change seems to have involved alterations to the melanocortin 1 receptor (MC1R) gene. Alan Rogers from the University of Utah and his group showed that in modern Africans this gene contains large numbers of silent mutations, mutations that must have built up since the gene was altered when hominins first became hairless. Assuming that the human population at this time was relatively small—in the thousands rather than millions—such mutations could only have built up slowly, so Rogers calculated that at least 1.2 million years must have elapsed since hominins lost their hair. It must have been our ancestor *Homo erectus* who first became hairless, at a time when it was walking bipedally, living in increasingly treeless savanna habitats, using fire to deter predators and cook food, and living in small communities.

But what drove hair loss? The generally accepted explanation among anthropologists is that losing hair allowed early humans to keep cool in the hot savanna regions into which they had moved. This idea has been widely expounded from the 1960s, most notably by Peter Wheeler of Liverpool John Moores University, and it is extremely attractive and superficially plausible. After all, we put on more clothes to keep warm and peel them off in hot weather to help us keep cool. Hair, like the clothes we wear, is a good insulator because it traps the warm air around our body. Losing our body hair would therefore enable us to lose heat more quickly via convection. It would also allow us to bring to bear another method of keeping cool—sweating—which cools us because energy is required to evaporate water from our skin. A naked body can remove more heat by this process than a hairy one because the water in sweat can evaporate more easily from the bare skin, rather than having to wet the whole pelt first. So though most hairy mammals, including primates, lose much of their heat by panting, which allows water to evaporate from the lining of the mouth, in humans sweating has taken over as the main form of evaporative cooling; it allows us to remove heat from all around our body and at several times the rate of other mammals.

So effective is heat removal by sweating that anthropologists have gone on to suggest that losing our hair was crucial for another advance in the evolution of humans: the ability to hunt large animals. Early humans would have found it difficult to kill large mammals such as antelope, wildebeests, and zebras on the African plains because of their inadequate physique. They would not have been fast enough to catch them or powerful enough to subdue them. The hunting hypothesis suggests that early men could run farther and for longer in the heat of the day than prey animals because sweating would keep them cooler for longer. If they tracked their prey for long enough, their prey would eventually overheat and become immobilized, allowing the hunters to catch up with them and dispatch them. The San Bushmen of the Kalahari are certainly known to be able to kill antelope using this technique, following them and tracking their movements for two to five hours, until their prey collapses; there is even a fine BBC video of a San Bushman tracking and killing a kudu in this way. However, it is not certain that it is hairlessness that gives the San Bushmen the advantage; two other mammalian predators also hunt in this way in savannas, African hunting dogs and spotted hyenas, and both of them are covered all over with hair like the prey they hunt. In fact endurance hunting is rare in hunter-gatherer societies, maybe because it has the disadvantage that though the hunter can keep cool by sweating, by doing so he loses large amounts of water (US army recruits have been known to lose over four quarts of water per hour when exercising in the desert!). The resulting dehydration can be fatal if weight losses exceed 2 percent of body weight. Nowadays hunters can carry water bottles with them to keep up their fluid levels, but there is no guarantee that early humans had invented vessels that were capable of carrying water.

But the hypothesis has a more fundamental problem, one rarely mentioned by anthropologists. In the heat of the day a naked body would actually absorb *more* heat than one covered in hair, meaning it would need to be more actively cooled. You might think that this would only occur when the air temperature exceeds our body temperature of 98.6°F, when heat would enter our bodies via convection. This only rarely happens in savannas, where the mean daytime maximum temperature is usually around 84°F. However, this leaves out the most important mode of heat transfer between our bodies and the environment: radiation. On a hot sunny day a hairless human body will absorb long-wave radiation emitted from the hot ground, and even more important, the much larger amount of short-wave radiation (mostly light) that comes from the sun. On such a day the net radiation entering our bodies can amount to around 670 watts per square yard, much more than the amount of energy we ourselves generate. The layers of hair follicles on a hairy animal will shield it from practically all of this radiation, so while the surface of its pelt might be hot, its skin remains at body temperature. For this reason, most savanna mammals are hairier than their cousins that live in dense forest and tend to have particularly dense hair on their upper flanks to ward off the sun's rays. Protected by their heavy fur coats, they have to use far less water to keep themselves cool than naked humans.

In deserts the problems of keeping cool in the daytime are most acute. It is noteworthy, then, that those "ships of the desert," the camels, have particularly heavy coats of hair on their upper flanks, while their human riders cover themselves with loose flowing robes. The shielding effect of hair also helps explain why humans have maintained a dense covering of hair on the tops of our heads; it helps us keep our most vital organ— our brains—cool. The importance of hair in thermoregulating our brain was driven home to many English cricket supporters back in 1994, when the English all-rounder Chris Lewis shaved his head at the start of a tour of the West Indies. He promptly went down with sunstroke, ironically missing the opening "warm-up" match. So important is our head hair in keeping our brains cool that the human races that inhabit hotter parts of the world, such as Native Americans and Africans, have lower rates of male-pattern baldness than the Caucasian inhabitants of the cool regions of Northern Europe. Presumably the disadvantage of losing the shielding effect of head hair is so great that baldness is strongly selected against.

And at least 50 percent of my readers may well have spotted another problematical aspect of the hunter hypothesis: its inbuilt sexism. The researchers who have investigated this theory (almost all men) have concentrated on an activity, hunting, that they have assumed was also carried out entirely by men. They totally ignored the contribution of women, who, they assume, spent much of their time "gathering" or perhaps simply waiting for the men to bring home their catch. They do not explain how hairlessness would have helped the women dig up roots, make fires, or cook. Indeed, according to the theory, women should actually be hairier than men since they would not have had such great cooling demands on their metabolism, whereas the reverse is true.

In recent years, therefore, several scientists have championed an alternative hypothesis, one that was first put forward in 1874 by the naturalist Thomas Belt, and one that applies to both sexes: that humans lost their hair to reduce their ectoparasite load. The reasoning is that hair loss occurred because early humans were now living and sleeping together in semipermanent camps, rather than in solitary nests. Ectoparasites would therefore be more likely to build up around the camp and become more of a problem. It is certainly true that before the advent of modern insecticides, we were highly troubled by such parasites. Our mattresses were infested by bedbugs, our head hair by lice, and pubic hair by crab lice. Moreover, humans are the only one of the 193 species of monkeys and apes to have its own species of flea, *Pulex irritans*, something that is only possible because we live in permanent settlements; the larvae fall to the floor and live on organic debris in our houses, guaranteed to find new humans to bite once they have emerged from their pupae as adults.

Ectoparasites are not only irritating and suck our blood; they also carry dangerous infectious diseases such as typhus, various forms of spotted fever, and bubonic plague. There would therefore have been strong selection pressure on any morphological feature in early humans that would have reduced the ectoparasites' numbers. The ectoparasite theory suggests that the best way to do this was to lose our body hair. This certainly fits with our historical experience: in World War I it was found that cutting soldiers' hair shorter greatly reduced the buildup of head lice. The "short back and sides" haircut subsequently became the fashionable norm for men right up until the 1960s. And reducing the length and thickness of our hair not only makes it easier to visually spot fleas and lice on our skin; recent research by Isabelle Dean and Michael Siva-Jothy of the University of Sheffield, England, has also shown that our fine body hairs act as excellent movement detectors, allowing us to feel where the parasites are. Finally, the theory also provides a satisfying explanation of why women are less hairy than men: staying longer at camp than the men, they may have been more prone to being loaded with parasites.

The ectoparasite hypothesis therefore provides what I think is the better explanation of why we lost our body hair, but whichever hypothesis you favor, the benefits would have had to be large enough to overcome a serious disadvantage of nakedness. Naked *Homo erectus* individuals would have suffered from a quite different thermoregulatory problem than overheating during the day: they would have been extremely prone to getting cold at night.

Because of our lack of insulating hair, modern humans are poorly adapted to coping with changes in temperature. All warm-blooded animals have a range of air temperatures at which they are comfortable and at which they can keep their core body temperature constant without having to raise their resting metabolism. Within these temperatures they can regulate their body heat merely by changing their behavior—by curling up, for instance, or stretching out. As you might expect from what I have outlined above, our upper critical temperature is quite low, around 97°F, even in deep shade, and our lower critical temperature is high, around 77°F. We could be comfortable living naked in a rain forest, therefore, where air temperatures range around 82°F–90°F (and rain forest tribes consequently tend to wear few clothes), but not elsewhere. In the savanna it also gets hot during the day: the mean maximum temperature in the Serengeti, for instance, ranges from 79°F–84°F. However, because the sky is often cloudless and the air is drier than in a rain forest, the mean nighttime temperatures fall to 57°F–61°F. What is worse, the clear night skies of the savannas are effectively 36°F cooler than the ground beneath them. Because radiation dominates heat transfer between ourselves and our surroundings, at least in calm conditions, we radiate a lot of body heat to the night sky (think how cold you feel at festivals or open-air plays after the sun goes down and how stiff your neck and shoulders get). So at night in the Serengeti it can effectively feel more like 43°F–50°F; tourists to the region are advised to bring sweaters and jackets for the cool evenings. A naked *Homo erectus* living 1.2 million years ago on the open plains of East Africa would therefore have got cold at night and have had disturbed sleep.

There are three possible ways out of this conundrum. Early humans could have huddled around the fires that they built and maintained overnight for protection from predators. Most of us have sat around campfires sometime in our youth, and they certainly warm the side that faces the fire. However, the side of our bodies that faces away from the fire and the top of our shoulders, which face the sky, can get cold. Out in the open our bodies also lose heat rapidly to the cold ground.

Another way they could have kept warm is to have used animal skins as bedclothes. However, it is difficult to believe that physical evolution could have been moving one way—making people colder—while behavioral evolution at the same time had to try to make up for it. Besides, the first actual physical evidence for clothes, or the tools such as needles needed to make them, comes far later in the story of humans—scraped hide three hundred thousand years ago, and sewn clothes just twenty thousand years ago.

It is far more likely that the *Homo erectus* were already doing something that would help keep them warm at night before they lost their hair; at their campsites they were already building shelters that helped protect them from the rain, shelters that would also have helped keep them warm. They would certainly have a good incentive to do this in the rainy season. None of the great apes like getting wet; Sumatran orangutans, for instance, often make second nests directly above their sleeping nests and use them as canopies to keep off the rain. For early humans, long used to building sleeping nests on which they rested, it would not have been a problem to construct simple huts to shelter themselves. Indeed, many tribes of hunter-gatherers still build small semipermanent huts from thin branches that they cut off savanna trees; they insert the thick ends of the branches into a ring of postholes in the ground and fasten them together at the top in the same way that apes weave their nests together. The frames are then covered with leaves or skins or even coated in mud. No huts have survived from the Paleolithic period to provide proof that early humans built them, but this is hardly surprising; the huts of modern hunter-gatherers fall apart within a few weeks or months of being abandoned and leave no trace. But tantalizing evidence of a 1.8-million-year-old building has been found at a human campsite, complete with fossil remains, at Olduvai Gorge, Tanzania. It comes in the form of a circle of stones, some thirteen feet in diameter, stones that could have helped stabilize and reinforce a circular hut or windbreak. The interpretation is controversial, and many anthropologists claim that the circle has been formed by natural means, but this stone circle may show the beginnings of human architecture.

You might think that flimsy wooden huts would provide little warmth, since cold air could rapidly penetrate such a drafty structure, but they can be quite effective, and anything that shields us from the cold night sky helps. Research on the climatic benefits of urban trees by my PhD student David Armson showed that people can feel up to 18°F cooler in the shade of a tree canopy, but at night they feel 2°F–4°F *warmer* because the leaves shield them from the cold night sky. Modern people such as the hunter-gatherers of the Hadza tribe of Tanzania still sleep beneath trees for this reason, if only during the dry season.

And we know that the benefits of sleeping inside simple huts is even greater than that of sleeping beneath trees because of a recent study by the sleep specialist David Samson and his colleagues. Being interested in understanding how humans became able to sleep for so long and so deeply at night, they investigated how several factors affect the sleep of the Hadza during the wet season, when they retire to rest at night in simple grass-covered huts built by the women. As well as measuring other factors that might affect sleep, such as noise, Samson and his team also estimated how warm people would feel both inside and outside their huts. They used simple weather stations that measured air temperature, relative humidity, and wind speed. Largely because the huts cut down air currents and shielded them from the cold night sky, Samson calculated that sleeping inside a hut would feel 8°F–10°F warmer than outside, enough to allow for a comfortable night's sleep.

It was therefore because early hominins were sleeping inside wooden huts that they could afford to lose their body hair. And this would in turn have made us even more dependent on our practical woodworking skills, to make fires and build ever-more-elaborate shelters, and eventually to use other materials to make sheets and clothing. Paradoxically, as we got better at these activities, we would have started to be able to colonize cooler climates. Becoming hairless forced us to become more ingenious and to rely on our intelligence to help us manipulate our environment, rather than have to adapt to it as other animals do. It would have helped a fairly feeble primate conquer the world.

CHAPTER 4

Tooling Up

F ew attractions are more enjoyable to visit for those of us fascinated by the world about us than local museums. When one visits a new town, there is nowhere better to get a feel for the place and a sense of its history: to understand why it is there and why it is laid out in the way it is. Charming and old-fashioned, the museums are a fine testament to civic pride and to the enthusiasm and curiosity of the local people. A treasure trove of everyday bygones—workmen's tools, sepia photographs of men in suits and bowler hats and women in voluminous skirts and aprons, model buildings and ships, stuffed animals and human skeletons—they bring our ancestors alive to us like nothing else.

The museums invariably start with a section on the local geography and geology, with a collection of fossils, before introducing a major display on "Xville in the early Stone Age." An imaginative diorama depicts Xville thousands of years ago. Early men, in skin loincloths are depicted holding spears and stalking mammoths or other large beasts, or returning to their camp bearing deer hanging from poles. Meanwhile "early women," dressed in rather less revealing skins, greet them from the campfire. And in front of the diorama are mahogany display cabinets with rows of stone tools. Arranged in age order, from what look to the uninitiated like merely random pebbles, to beautiful tear-shaped hand axes, finely flaked arrowheads, and smoothly polished ax heads, they present to the visitor the development of human stone tools that will be familiar to most students of prehistory.

The study of stone tools has dominated anthropology and archaeology ever since 1831, when the Danish antiquarian Christian Thomsen introduced the concept of classifying the "ages of man" according to their dominant materials—stone, bronze, and iron. His ideas were developed and popularized by the British baron John Lubbock. In his 1865 book, *Prehistoric Times*, Lubbock went on to split the Stone Age into the Paleolithic and the Neolithic, with the more contentious intermediate Mesolithic period in Europe being defined only much later. Since then, archaeologists have spent huge amounts of time and effort classifying stone tools, arranging them in chronological order, replicating their manufacture and use, and following their development. In doing so, they cemented in place a worldview in which the lives of our early ancestors, and in particular their material culture, was dominated by their relationship with stone. It was generally assumed that early "Stone Age" men were the first to produce tools; that the first tools were made of stone; that stone tools dominated their world; and that the sophistication of early stone tools demonstrated the mental superiority of early humans.

These assumptions seemed perfectly valid in the nineteenth century, and for the first half of the twentieth. After all, stone tools were the only human artifacts that appeared to have survived from the time of early hominins; anything made of organic materials—skin, plant fibers, or wood—had long since vanished. However, in the last fifty years new discoveries by primatologists and anthropologists mean that we now know that none of the assumptions made by nineteenth-century archaeologists are valid. First of all, primatologists have found that our relatives, the apes, produce a wide range of tools, so humans cannot be exalted over other animals because they were the first toolmakers. Second, most of the tools made by apes—spears, chisels, digging sticks, and nests—are made of wood, not stone, and it is highly likely that early hominins would have inherited their woodworking skills from the apes. So the first tools used by early hominins would have been made of wood, not stone. Third, even the reconstructions of the lives of early hominins that have been made by devotees of stone make it obvious that they used mainly wooden tools—to hunt animals, to dig up plant roots, and to construct shelters—and that they burned wood to keep off predators, keep themselves warm, and cook their food. If we cast our mind back to those dioramas in local museums, for instance, most of the tools they depict were actually made of wood. The men had wooden spears to kill game and used wooden poles to hang it from, and back at camp the women were standing beside wooden huts and cooking their food over wood fires. Stone tools were only used to butcher the animals that had already been killed and to scrape their hides to make skins.

Finally, in contrast with the claims of archaeologists, the first stone tools were hardly sophisticated objects, particularly if we compare them with the artfully constructed nests of apes. The earliest ones, the Oldowan tools, which date from 3.2–2.5 million years ago, often resemble random pebbles, and even the flakes produced by the Acheulean technology, which emerged 2.2 million years ago, are pretty rough and ready. After all they were produced rapidly, simply by hammering two lumps of stone together, or by hitting a piece of stone with a bone or a log of wood. Hand axes, which were first produced around 2 million years ago, certainly look more impressive and show evidence for the first time of clear design. However, even hand axes can be made in as little as twenty minutes and are essentially just tear-shaped flakes of rock with two edges. Their design remained largely unchanged for hundreds of thousands of years, so their manufacture demonstrates little evidence of intellectual progress. Only much later, with the sophisticated retouching techniques that were developed in the Upper Paleolithic period, around one hundred thousand years ago, did stone tools become sophisticated enough to impress any small child that we might have brought into the museum. Only then did humans shape blades that actually look like modern daggers, harpoons, and barbed arrowheads.

So stone tools were by no means as novel or central to the life of early humans as has been assumed. However, in any branch of learning, once a culture is established, it seems to be hard for those initiated into it to break free. Anthropologists have continued to this day to overemphasize the importance of stone tools and ignore those made from other materials. For instance, in his otherwise admirably clear book *The Origin of Humankind*, Richard Leakey includes a particularly illuminating passage that evokes a day in the life of a band of *Homo erectus*. In it, the women use "stout sticks" to dig for tubers, but the young girls seem to spend much of their time practicing making stone tools. The men, meanwhile, kill an antelope by unconvincingly striking it a "stunning blow" with a rock, before stabbing it with a "short pointed stick." Leakey never mentions wood. Indeed, in common with most books on human evolution, the word *wood* is not even deemed important enough to be included in the index. To see why early humans actually hardly ever used stone tools, why these tools were relatively small, and why they were used almost exclusively to cut things up, we need to compare and contrast the structure and mechanical properties of stone and wood.

The properties of stones result from their composition; they are made up of crystals or amorphous blocks of inorganic chemicals. In typical igneous rocks, such as granite and dolomite, these have solidified from a molten state, while in flint they have precipitated out from solution. Sedimentary rocks, such sandstones and shales, are composed of bits of igneous rock that have been pressed together, while chalk and limestone are made from the fossilized inorganic skeletons of dead organisms. The strong bonds between the atoms make stone extremely stiff and hard. This makes it ideal as an impact tool. If you use a stone to strike a nut or hit a piece of bone, it's the nut or the bone that will deform more, and all the kinetic energy in the stone will be used to break them. None will be absorbed by the stone. However, if two stones are hit together, the energy has nowhere else to go, and cracks will readily run through or between the crystals, breaking one or both stones. Stone is brittle and breaks easily, and if there are no predetermined lines of weakness, as in flint, hitting the stones together in the right way can create fractures in a predicted direction, creating sharp edges. The hardness of stone makes these edges ideal for cutting; they can withstand the large compressive stresses set up as the sharp point is pressed into or slid across a softer material such as flesh or even bone and slice through it. This is why sharp flint tools are ideally suited to butchering animals and scraping skins.

The brittleness of stone has a major downside, though. It

makes the material weak in tension, since small surface cracks can readily run through the whole stone; rods of stone, just like sticks of blackboard chalk, are easily snapped. Stone knives therefore need to be short and thick to prevent their blades from being loaded in tension, and even if a stone spear could be fashioned, it would be far too delicate to use; it would fall apart at its first throw.

In contrast, we have already seen that wood has evolved in trees to be strong in both compression and tension, and extremely tough along the grain, which is why tree trunks and branches are so good at resisting bending. Dried wooden branches have even better properties, being just as strong and tough as green wood, and three times as stiff. They are, therefore, ideal for making digging sticks and spears: they are rigid and strong in bending, so they don't flex or break when subjected to bending force; they are tough enough to withstand impacts; and they are still hard enough to pierce skin or soil. They are also relatively easy to make; they can be shaped when the wood is green, when it is still soft enough to be cut, carved, and finished.

It's thoroughly predictable, therefore, that most of the large tools that the early hominins used were made of wood and only the small cutting tools were made of stone. Their huts would essentially have been inverted versions of the nests built by their ape cousins, and their spears and digging sticks would have been similar to those created and used by savanna chimps. And there was probably little difference in the planning involved in producing wooden and stone tools. The tools that modern apes create are made for the moment and used immediately, either to sleep, dig, or hunt, and they are hardly modified at all from branches and twigs. To make a spear, for instance, the apes strip off the leaves and side branches with their hands and sharpen the thin end with their teeth, but otherwise it is used "as found." Their spears are irregular and bent, and the making of them shows little planning or forethought. The same could be said for the stone knives and scrapers used by early humans to process animal carcasses. They were made and used on the spot—so the fact that early humans made stone tools, or the details of the methods they used to make them, hardly shows a step change in intelligence. The increasing success of humans is best explained by their development of their wooden tools, particularly their weapons. These developments can also be used to chart the evolution of human intelligence.

The first real intellectual advance that hominins made must have occurred when our ancestors started to use stone tools not just to process their kills but to construct wooden tools. This would have had to happen when early hominins moved onto the savannas. They would have needed to make thicker digging sticks to get at roots and tubers in the dry season, and larger spears to hunt game bigger than bush babies. And they would have had to use larger branches to construct the huts in which they sheltered. The first fully terrestrial hominin, *Homo erectus*, would not have been able to do this without using tools. With their small incisors, they would not have been able to sharpen their spears and digging sticks, and having less powerful arms than their arboreal ancestors, they would not have been able to break off big enough branches to make their shelters. They would have needed to use stone scrapers to sharpen the points of their tools, and to use stone knives, axes, or saws to cut off branches. *Homo erectus* would have had to become the world's first carpenters. In doing so they would be the first primates to make a tool not just for immediate use, but to make another tool. The first evidence of carpentry has indeed been found shortly after *Homo erectus* appeared. In 1981 Lawrence Keeley and Nicholas Toth of the University of Illinois found polishes typical of those left by woodworking on 1.5-million-year-old stone tools from the Koobi Fora region of Kenya. And in 2001 Manuel Dominguez-Rodrigo and coworkers from the University of Madrid discovered calcium oxalate crystals from *Acacia* wood attached around the blades of flakes and hand axes from the 1.6-million-year-old site of Peninj in Tanzania. This showed that the hand axes had been used for woodworking, something that was confirmed by wear patterns on the tools that indicated heavy-duty activities such as sawing through branches.

It may not seem much of an advance, but using stone tools to carve spears involved a step change in the human imagination and shows major progress in intellectual and social organization. Miriam Haidle, from the University of Tübingen, has pointed out the huge differences between two activities that at first glance might seem similar: the manufacture of spears by chimpanzees and of similar tools by early humans. The chimp fashions every aspect of its spear where and when it will be used. It strips off the leaves and side branches of the branch with its hands and sharpens the thin end with its teeth. When hominins made a spear using a hand ax, their actual actions are not necessarily more complex, but the process did involve two separate sets of actions that could take place at different times and places: making a hand ax, and then using it to make the spear. The whole process therefore involved not only integrating information from the past, using so-called working memory, but also imagining future actions, using what has been called constructive memory. Haidle analyzed the two processes by classifying and counting all the steps needed to make a spear, and the objects on which they are focused, to produce what she called a cognigram. Making the chimp's spear involves fourteen steps, which acted on three "foci," the chimp itself, the prey item, and the tool. In contrast, the human spear took twenty-nine steps, acting on eight foci. The complexity of the task had been more than doubled.

The two processes of hand ax and spear manufacture need not have been carried out by the same individual. Just as I find my Swiss Army knife (made by other people) invaluable and carry it around with me everywhere, early humans may have carried around hand axes that were made by someone else. The process might, therefore, also show evidence of greater social organization, not just better individual mental capacity, within *Homo erectus*. But the great advantage of analyses such as that of Haidle is that they enable us to follow the development of intelligence and social cohesion within early humans even in the absence of other evidence. Simply by analyzing the subsequent advances in toolmaking, we can use objects to infer the evolution of intelligence.

The main difficulty in following the progress of human intelligence in this way, though, is the lack of evidence in the form of the wooden tools themselves. We have found no actual wooden objects for the first million years following those first signs of woodworking, so we do not know what tools *Homo erectus* made. This has led many anthropologists to doubt the importance of wooden tools, and in particular to doubt the hunting capability of these early humans. It is often thought that hominins might until recently have been at best opportunistic scavengers, only able to rob the carcasses of large herbivores, maybe acting together with small stabbing spears to drive away other carnivores from the prize. However, recent years have brought exciting finds of wooden spears that are starting to revolutionize our ideas about our early ancestors.

Only when humans colonized the cooler, wetter parts of the globe did conditions allow wooden tools to be preserved. One of the main reasons we have such a fine archaeological record of early humans in Europe is that the wet, acidic peat soils that accumulate in the colder regions protect organic materials such as wood from rotting and preserve them surprisingly intact. I remember to this day the horror as an eight-year-old of seeing the perfectly preserved corpse of Tollund Man in the Silkeborg Museum, near Aarhus, Denmark. His face, in particular, looked almost alive, down to the stubble on his chin. A victim of ritual sacrifice over two thousand years ago, he gave me nightmares for months afterward, and troubling intimations of mortality.

The earliest recorded wooden tool is the Clacton Spear, which was uncovered back in 1911 by the amateur prehistorian Samuel Hazzledine Warren, in 450,000-year-old deposits in Essex, England. It provided a first tantalizing glimpse of the woodworking capability of early humans. The "spear," actually a sixteen-inch-long fragment of yew, is the pointed end broken off a larger object that has variously been interpreted as a digging stick, a lance, or a spear. Close examination reveals that it tapers evenly to a sharp point, making the last interpretation most likely. Modern experiments to replicate how it was made, which were carried out by John McNabb and Hannah Fluck of the University of Southampton, England, suggest that it was shaped by scraping it toward its tip using one of the large blade tools found at the site, a Clactonian notch. If the process had been carried out on seasoned wood, it would have taken up to two hours, so the spear may well have been shaped while the wood was still green. McNabb suggested that the carver might instead have used fire to help shape it. The end could have been held in the embers of a fire to heat it up and char the outside, which could then be scraped off. Using such a process, Fluck found that the tip could have been shaped in as little as forty-five minutes. As they pointed out, this might have had the added advantage of "fire hardening" the tip. To test this idea, I set an undergraduate project student, Michael Chan, to investigate the effect that fire hardening might have. He compared the mechanical properties of coppice rods of hazel, one-half of which had been heated over a disposable barbecue, and the other half just left to season slowly. He found that the heating did harden the wood by a small amount—which is what you would expect as it would crystallize the hemicelluloses in the cell walls—but it also halved the toughness of the wood. This could have been the reason why the tip of the Clacton Spear had broken off and suggests that fire hardening of wooden tools is not particularly useful; it just seems as if it should be, as if the fire is transferring some of its strength to the tool, and gives people more faith in the power of their weapons.

The discovery that totally revolutionized our understanding of early hunting techniques and the life of early humans was not made until the 1990s. From 1982 on, Hartmut Thieme of the heritage office of Lower Saxony, Germany, had been carrying out a long-term operation to rescue Middle Paleolithic sites threatened by large-scale open-cast lignite mining at Schöningen near Hannover. In the fall of 1995, excavating the former shore of a lake, Thieme and his colleagues were astonished to uncover a cache of beautifully carved wooden implements, along with the slaughtered bodies of over twenty horses. Seven of the implements were clearly identifiable as throwing spears. Carved out of the narrow trunks of slow-growing spruce saplings, they were between 5 and 7 feet long and 1.2 to 1.6 inches thick, tapered at both ends and with the widest point a third of the way along. They closely resembled modern Olympic javelins; subsequent experiments on replicas have shown that they would have been stable in flight and could have been thrown accurately up to sixty-five feet. Impact wounds on the horses' skeletons suggested that they had been killed by being hit with these weapons.

The sheer number of spears and corpses found by Thieme suggested that the site must have been an ambush area; the early humans, who would have belonged either to the species *Homo heidelbergensis* or to our even closer relative *Homo neanderthalensis*, must have acted together in a group to cut off the horses between dry land and water before slaughtering them, though the horses had probably not all been killed at the same time. Altogether, the finds speak volumes of the sophistication of these early humans. They were not only capable of fine carving, of being able to imagine the shape of the spears within the trees and shape them with stone tools, but were also able to organize themselves into efficient hunting parties, to exploit the behavior of their prey animals, and kill them safely from a distance.

Some of the wooden tools at Schöningen also showed signs of hafting—grooves around the tip suggested that small flakes of stone had been incorporated in and around the tips of several tools, apparently to make them more effective cutting weapons. Indeed, such "composite tools," combining the strength and rigidity of wooden handles with the sharp cutting edges of stone, seem to have appeared elsewhere at this time, including at many sites throughout Africa, thus marking the beginning of the Middle Paleolithic period. The archetypal composite tool that was developed was the stone-tipped spear, the sort of weapon beloved by Hollywood, and which early humans are so often depicted brandishing threateningly in caveman movies. Rather than relying on a sharp wooden point, Neanderthals and early *Homo sapiens* started to haft a sharp stone tool, rather like a hand ax, to the front of their spears, cutting a groove in the end to receive it, and holding the blade in place using a combination of animal glue and sinew binding. The manufacture of composite spears was therefore extremely complex, with several separate tasks or "modules"—preparing the rope; boiling up the glue; shaping the stone point; and cutting the groove in the handle—even before the final assembly. This shows even greater organizational and technical ability and intelligence on the part of the Neanderthals. I find it hard to imagine that I would be able to carry out such a complex task without lengthy training.

The puzzling thing, though, is that it is hard to see the advantages of using such a composite spear. Many anthropologists have carried out experimental investigations comparing the killing performance of stone-tipped and simple wooden spears. The experiments, carried out on the corpses of pigs or on ballistic gel, must have been great fun to perform. However, though the experimenters were clearly expecting the stone-tipped spears to be better at penetrating flesh, they found little evidence of this. Both wood and stone are harder than skin, so they both cut through it with ease. In some studies, the wooden tips even penetrated farther than the stone ones, though there was some evidence that the wider stone blades could cause damage over a greater volume of flesh. They might have been better at spilling blood. But composite spears have the disadvantage that the brittle stone tip is more prone to snapping off, so they would need mending more often. The real advantage may have been due to the higher density of the stone. The heavy tip of the spear would bring its center of gravity forward, even if it was just mounted on a short cylindrical handle. This would enable it to be thrown effectively, while it could also be held and used as a stabbing spear. Composite spears could therefore act both at close quarters and at a distance and be used as both offensive and defensive weapons.

But both wooden javelins and composite spears have a limited killing range. The shortness of our arms means that we need to contract our arm muscles much faster than at their optimal speed to move the hand holding the spear forward and upward. Furthermore, of all the energy used to accelerate our arms and hands, around half is wasted. This limits the speed we can impart to a hand-thrown object, so few people can throw spears of any type more than thirty yards. Fortunately, though, our ancestors developed several ways to overcome this problem and make themselves into more efficient hunters; and most of them did so using techniques that worked by artificially extending the length of their arms.

Before the pits of the South Yorkshire coalfield were shut down by Margaret Thatcher in the 1980s, one of the hobbies of the young miners was the unlikely pursuit of arrow throwing. They wrapped the end of a piece of string one turn around the back of an arrow shaft and looped the other end around their index finger, holding the string taut while also holding the arrow farther forward along the shaft. They then threw the arrow, releasing the shaft with their fingers early on, and allowing the string to continue accelerating the arrow until it unwound from the shaft and released it. The string effectively lengthened the miners' arms, allowing them to put more energy into the throw and to propel arrows two hundred yards or more!

It would be difficult to use exactly the same technique to throw a spear because the force of the string on a single finger would be too great. A similar method was, however, used by the ancient Greeks. Their lightly armed peltast troops used lighter javelins than the traditional hoplites and could throw them farther; they threw their javelins with the aid of leather thongs called amenta, which they looped over two fingers. It is also becoming clear that from around twenty-three thousand years ago Upper Paleolithic *Homo sapiens* did much the same, but using a special tool to hold the string. From the beginning of the twentieth century, archaeologists had been unearthing decorated rods of wood or antler into which a hole had been drilled toward the wider end. At a loss to describe their purpose, the archaeologists at first fell back on their default position: they assumed that the rods were used for "ritual" purposes, and, noting their similarity to scepters, gave them the name *bâtons de commandement*. Later investigators suggested they were used to straighten spears, which could be held within the hole in the stick. However, there is no reason to believe that such tools would be needed since spears can easily be straightened without them. Fortunately, other more practical people in the world had an interest in these objects. These are the amateur enthusiasts of "primitive technology," mostly men with impressive tool kits and engineering aptitude who carry out experiments in their own homes and sheds, and who self-publish their findings or nowadays make fascinating YouTube videos. They have shown that if a string is attached through the hole in the baton, and the long end of the baton is held in the hand, it can be used to propel a small spear or dart in the same way as the miners' arrows, extending its range to around sixty yards. Further evidence that this was indeed how they were used was found on a fragment of a *bâton de commandement* unearthed at the famous cave of la Madeleine, Tursac, France. A simple carving shows a man holding a dart just as he would if he was about to throw it with a *bâton*.

Even better results can be achieved to throw a spear farther and with greater accuracy by using a spear thrower. Also developed in the Upper Paleolithic and still used extensively in Central and South America (where they are called atlatls) and in Australia (where they are known as woomeras), the spear thrower is a simple stick six to eighteen inches long with a cup or hook at the far end. To use it, the thrower is held horizontally under the spear, its hook overlapping the back of the shaft, while the hand holds on to the shaft farther forward. The thrower acts as a third joint to the arm, the spear or dart being propelled forward by rotating the thrower forward with the wrist at the same time as the arm. The mechanics is identical to that of the modern dog-ball throwers that allow pet owners to exercise their dog with minimum exertion to themselves. I used this fact to safely investigate how the effectiveness of atlatls depended on their length. My master's student Hannah Taylor filmed how far her family and friends could throw balls of different weights using an adjustable dog-ball thrower. She found that for light balls, long throwers were best, but the optimal length fell as the weight of the ball increased; people simply did not have strong enough wrists to propel heavy balls forward using a long lever. Many people still enjoy using real atlatls and compete under the auspices of the World Atlatl Association.

According to Wikipedia, the world-record distance for a throw using an atlatl is an amazing 848.56 feet, set by Dave Ingvall of Missouri, on July, 15, 1995.

Yet another technique to increase the killing range of wooden tools was to use the stick itself as an extension to the arm and rotate it forward as it was thrown, like a person throwing a stick for a dog. This technique is fairly effective at increasing the speed of the stick when it is released, but as the stick tumbles through the air, it slows down far faster than a spear because of the increased aerodynamic drag. These problems were overcome, though, by the people who perfected this method, the Aborigines of Australia. They invented a wide range of boomerangs, all of which have a streamlined cross section to reduce drag and help them fly through the air. Some (the less crooked ones) are designed to fly straight and can be lethal at up to two hundred yards, while the more bent ones are designed to curve through the air in flight and so return to their owners.

But of all the ways of improving the killing performance of wooden projectiles, the best is the bow and arrow. This combination was probably first invented some sixty-five thousand years ago in Africa, though evidence from Europe only seems to go back some twenty thousand years. Rather than relying on the fast-twitch performance of our arm and shoulder muscles, bows make use of the larger forces and greater energy we can produce when these muscles contract slowly. As we pull back on the string, elastic energy is stored in the bow, which is subsequently released when we let go of the string, propelling the arrow forward. Much has been written about the complexity of the mechanics of bows and why they are so efficient, but the process is basically simple. When a bow is fully drawn, the string is angled back sharply, so when it is released, some of the energy is used to accelerate the arms of the bow forward as well as accelerate the arrow. As the string straightens, however, trigonometry dictates that even a tiny movement of the bows' arms results in a large movement of the arrow. By the time the arrow is released, the arms of the bow have effectively come to rest, having transmitted virtually all of their elastic and kinetic energy to the arrow.

Bows have three major advantages over all the other techniques we have seen. First, since our muscles can produce more energy when contracting slowly, a bow can release more energy to a projectile, so that arrows can be shot over nine hundred feet. (According to Wikipedia the farthest accurate shot in archery under World Archery conditions is actually 283.47 meters [930.02 feet], achieved by Matt Stutzman [USA] at the TPC Craig Ranch, McKinney, Texas, on December 9, 2015.) Second, since a bow is drawn with a slow, smooth movement, it can be aimed far better and is a far more accurate weapon than a spear. Finally, since from the front the archer barely seems to move, she or he is far less conspicuous to prey than a javelin thrower, so the bow and arrow makes a much better stealth weapon. The bow and arrow quickly became the hunting weapon of choice, not only for hunters on the savannas and prairies, but also those colonizing the dense forests that sprang up in Mesolithic times at the end of the last ice age.

Bows are extremely effective, but they are vastly more complex tools to manufacture than composite spears. Marlize Lombard and Miriam Haidle have calculated that it takes 102 tasks, spread across 10 subassemblies, to make a complete bow and arrow set. In their hunting technology humans had indeed come a long way from the simple stabbing spear of the chimpanzee to the hafted and fletched arrows, and gut-stringed bow of modern humans. Our killing power had extended from the simple close-quarters dispatch of a small primate, to being able to kill large ungulates (and other people) over two hundred yards away. The development of wooden weapons had made us an apex predator, allowing us to inflict a mass extinction on the world around us. Even before we had learned to modify our environment by farming it, we had used wooden tools to kill off such magnificent beasts as mammoths, woolly rhinoceroses, and giant elk in Europe, giant orangutans in Asia, mastodons, horses, and tapirs in North America, giant ground sloths and armadillos in South America, and giant wombats and kangaroos in Australia. And the English victories against the French at Crécy and Agincourt show that the wooden bow in its ultimate development as the yew longbow was still the most effective weapon of mass destruction right up until the fifteenth century.

PART 2

BUILDING

CIVILIZATION

CHAPTER 5

Clearing the Forest

If there is one symbol of the Neolithic period, the time when humans first made a major impact on the environment by starting to farm the land, it is the polished stone ax. Indeed a major recent account of the Neolithic in Europe was called *The Tale of the Axe*. You can see polished ax heads in local museums around the world, and if you get a chance to handle one, then take it. Beautifully smoothed and rounded by grinding and polishing, they weigh heavily in the hand, into which they fit snugly. The expanded blade end tapers to a smooth if not sharp edge, while the back curves into a blunt dome. We now know what these ax heads were used for, but when they first started to be dug up and came to the attention of antiquarians, people had no idea what they were. This should not surprise us; after all, they don't look much like heads of modern axes, which are far narrower with far sharper blades. In fact it does not look as if they would be much use for cutting at all. The farmers who found them buried in their fields called them thunderbolts, imagining that they had been thrown down to the ground by some deity during storms. Antiquarians quickly came to view them instead as ceremonial objects, and it seems that some of the most slender ax blades were used as offerings to the gods;

pristine specimens were often found buried as grave goods in Neolithic long barrows.

Only in the last sixty years have we started to realize how effective Neolithic polished axes could be at cutting wood and what a vital role they played in our rise to civilization, clearing the forest, spreading farming around the world, and building the first farmsteads, villages, and towns. As we shall see in this chapter, their success is just the first of many cases in which technological advances in other materials helped people make far better use of the material they had always exploited: wood.

But polished stone axes might never have been developed had the world's climate not begun to change around fifteen thousand years ago. We saw in the last chapter that by then the one remaining human species, *Homo sapiens*, had perfected their hunting technology, enabling them to tackle even the largest of the beasts that roamed the plains and tundra during the last ice age. But as the climate became warmer and wetter at the start of our present interglacial, forests started to advance. Throughout the northern hemisphere, in Europe, Asia, and North America, people had to modify their weapons to deal with the smaller browsing animals that thrived in such forests: deer, wild cattle, and wild boar. Instead of heavy stone blades they attached smaller, sharper flakes of flint known as microliths to their darts and hafted finely carved barbed heads onto their arrows. But they also needed to develop tools to cut through the trunks and branches of trees so that they could open up small clearings in the forest where the fresh regrowth would attract game and where they could build their camps. They could have cut down small saplings by sawing through their stems with serrated stone blades, just as the North American Indians did until recent times. But sawing wood with such crude tools is a lengthy, inefficient process, and impracticable for trunks over an inch or so thick; it would just take too long. To cut down the larger trees that were growing farther and farther north, the Mesolithic people of Europe hafted small flaked flint blades into wooden handles to produce what are known as tranchet axes. The same was true in the Americas, though the Dalton people of the Mississippi also produced basalt-bladed adzes, tools that are similar to axes except that the blade is orientated at right angles to the handle, not parallel to it. Just like spear throwers, the handle of these axes and adzes effectively lengthened the arms of the lumberjacks, enabling them to impart more energy to the head of the ax. To cut down trees, they had to swing the blade down at an angle to the trunk, splitting the wood and cutting obliquely through the trunk. Repeated blows around the first cut could then take off shave after shave of wood, cutting a broad wedge through the trunk. Eventually, if they repeated the process all around the trunk, they could reduce it to a pencil-like point, like the pattern made by foraging beavers, and eventually sever the trunk.

The ability to fell trees precipitated the rise of a whole new material culture, one that took off across most of the forested regions of the world, but which has been best studied in Mesolithic Europe. Archaeological investigations have shown it enabled Mesolithic people to build roomy round houses. For instance, excavations led by Clive Waddington in 2002 at Howick near the Northumbrian coast, unearthed the impression made by a ring of post holes. These were the remains of a circular hut some twenty feet in diameter that dated from around 7600 BC. A reconstruction, funded by the BBC, resembled a tepee from the outside, but the structure was actually more complex. Short pine logs were buried into the postholes, and these were joined at their top by a ring of logs. This acted as a support for long, slender birch poles that rested on the earth at one end, were tied to the ring in their middle, and rose to a point above the center of the hut. Finally the poles were lashed together with smaller branches and covered in turf. It was a design to which we shall see people have repeatedly returned. Remains of an even earlier hut, dating from around 9000 BC, were found in 2008 at the famous Mesolithic camp at Star Carr, North Yorkshire, England.

Once a tree was felled, the trunk could also be split longitudinally to make thinner, more usable beams and planks. The easiest way to split wood is radially into pie-shaped segments, as in this direction the crack runs between the rays and through the weak pith at the center of the trunk. Whole trunks can be split remarkably easily, and using little energy, by simply inserting wooden wedges at the ends and sides and hammering them in to open up a crack and eventually split the trunk apart. At Starr Carr, people had made a pathway to their lakeside using a line of split planks, placed flat-side up. The two halves of a log can also be cut into quarters in the same way, and then into smaller and smaller pie slices. In most tree species, the trunk can also be cut into planks by splitting it tangentially, though this process is rather trickier to perform since it involves cutting through the ray cells, and it takes more energy. However, in 2007 a yard-long piece of oak that had been split tangentially was found in eight-thousand-year-old sediments at Bouldnor Cliff on the Isle of Wight, UK, suggesting that Mesolithic people had mastered this technique thousands of years before archaeologists had thought it possible.

The new woodworking technology also allowed people to

improve their mobility and hunting prowess by constructing two very different types of boat. The evidence suggests that the first watercraft that were developed in the northern forests were wooden-framed craft that were covered in animal skins. They were used by reindeer and caribou hunters who were being forced northward along with their prey into Scandinavia, Siberia, and Canada. No complete boat has survived, but open skin boats were pictured in later Norwegian rock carvings. A carving from Evenshus on Trondheimsfjord shows a hunter and his catch, while one from Kvalshund on the Repparfjord shows two hunters in a boat hunting a swimming reindeer. Earlier evidence of the way the Mesolithic people hunted reindeer and the remains of part of an actual boat have been found farther south, in Germany. At a ten-thousand- to eleven-thousand-year-old site in Ahrensburg, northeast of Hamburg, a reindeer skull was found with a hole in its forehead; this had been caused by a blow from one of the antler hatchets that was found at the same site. The hunters who killed it could only have approached this close to such a powerful animal if they had paddled up to it in a boat when it was swimming across a stretch of water. This technique, which is still practiced by the North American Inuit to kill caribou, would have been used to intercept deer migrations and lay in a supply of meat that could have been preserved by air-drying or smoking it. The remains of part of the frame of an actual boat from the ninth millennium BC was unearthed at Husum in Schleswig-Holstein, in the form of a curved section of an antler. From this fragment, experts at the German Maritime Museum, Bremerhaven, reconstructed the boat. They used antler frame members like the one that had been found and joined them to a wooden keel made of a branched piece of birch and long strips of birch to support the sides in just the way that present-day Inuit kayaks are built. Finally they stitched skins around it using bone needles to produce a fast lightweight craft—the world's oldest boat.

A quite different sort of boat, the dugout canoe, was developed by people farther south, the Mesolithic people of lowland Europe and the Dalton people of the Mississippi, who found a way to exploit the large trees that were now growing there. Having cut down a tree, they then hollowed the trunk out and used it as a boat. Of course it was not as simple as that. They had to remove large amounts of wood from the center of the trunk, and to do this they probably used fire, just as Native Americans still did even in recent times. The fire weakened the wood by charring it, after which it could more easily be removed with axes and adzes. The earliest log boat yet discovered was dug up near Pesse, in the Netherlands, and was dated at 6300 BC. It was still small, being only three yards long and cut from a pine tree with a diameter of just eighteen inches. It must have been suitable only for a single person. But log boats were probably fairly common at this time. The split planks at Bouldnor Cliff seem to have been just one product of a Mesolithic boatyard. And larger log boats have been found from later sites throughout Europe, and the technology to build them must have developed rapidly. Indeed, by the fourth millennium BC, log boat builders had finessed the design to make a craft from several components; the thirty-three-foot-long and twenty-six-inch-wide log boat found at Tybrind Vig, Denmark, has the rear end of its limewood hull reinforced and made watertight by an inset panel or transom across its stern. Log boats must have been common all around the world, from the Americas to Africa and Southeast Asia, and have in places remained the main form of transport right up until modern times.

In Europe there is evidence that skin boats and log boats enabled people to trade goods long distances; finds of items far from their sites of origin have been made along such major waterways as the Rhine and its tributaries and show that at even this early stage wooden boats were enabling long-distance trade and revolutionizing society. And the evidence from sites such as Star Carr and the Dalton sites of Illinois suggests that people were settling down. They were trading goods rather than shifting camp.

If reforestation was leading people to alter their way of life in northern temperate regions, elsewhere in the world climatic warming and wetting led people to make an even more drastic change and take the single most important step in the march toward civilization. They abandoned their hunter-gatherer lifestyle and become farmers. The first place where this occurred was in Southwest Asia, most notably in the hills of Anatolia, southern Turkey. A climate developed there in which warm, wet springs were interspersed with hot, arid summers and cold, freezing winters. Trees could not survive this combination of seasonal drought and cold, and these conditions instead favored fast-growing annual plants that could germinate in early spring, grow rapidly, and put all their energy into seeds by the end of summer. Since annuals do not have to lay down woody tissue like trees or store sugars in their roots as perennial herbs do, annuals are much more productive; they can produce far more fruit and seeds. People quickly settled down to cultivate and harvest annual grasses—the ancestors of our barley and wheat—for energy, and annual legumes such as lentils, chickpeas, and peas for protein, and became farmers. They harvested their crops using simple stone sickles and prepared the seedbed using wooden digging sticks similar to those used by hunter-gatherers, and mattocks, similar to the adzes of the Dalton people, but they had no need of more sophisticated tools until farming spread farther afield. At first it moved south into the Fertile Crescent, and in particular into the seasonal wetland surrounding the delta of the Euphrates and Tigris Rivers. Here, as along the banks of the Nile River, fresh mud was uncovered by the retreating floodwaters each year and formed a ready-made seedbed. Later, as the farmers became more successful, they moved northwest above the Tigris and Euphrates valleys where the land was slightly more arid. The people must have employed wooden spades to dig irrigation canals and used wooden buckets and other lifting devices such as the shadoof to raise water and carry it to their cultivation plots. The lands of the Fertile Crescent quickly became the bread baskets for the world's first civilizations and gave rise to the first large settlements such as the biblical cities of Ur and Jericho.

Arable farming could not easily spread eastward, being hindered by the cold, dry winters and summer drought of the central Asian steppes, where only perennial grasses would thrive. Here the land became dominated by pastoralists who herded sheep, goats, and horses. But farther west there were more promising territories for the spread of arable crops. The land around the Mediterranean, with its mild winters and warm, wet springs, provided ideal conditions for the growth of annual crop plants, while farther north and west, in central and Northern Europe, the warm, wet summers also provided an excellent growing season. The only problem was that in these regions the conditions were also ideal for the growth of trees: evergreen broad-leaved trees such as oaks and carobs around the Mediterranean, and deciduous oaks, beeches, ashes, and limes in Northern Europe. The land would have to be cleared before people could grow their crops. This was not so difficult around the Mediterranean. Here, the aromatic chemicals that helped the leaves counter summer drought made the trees more flammable, so the people could clear the ground by burning the vegetation during the dry summer months. Farming consequently spread relatively quickly through the lowlands of Greece, southern Italy, and right across to Spain.

Clearing trees was far more difficult in central and Western Europe, though, where the wetter weather and larger size of the trees made them much harder to burn. In these regions the people would have had to clear the trees by cutting them down, though like American Indians they could have ringbarked them first to kill them and lit fires around their trunks to make felling easier. And they could have let cattle or pigs loose to browse off the resprouting shoots. Tranchet axes would have proved inadequate to fell the trees, however. Though the axes are sharp, they have a rough surface; they are therefore apt to get stuck in timber and are prone to shattering because the impact stresses get concentrated around the surface ridges. They were also simply too small and light to cut deeply enough into a thick tree trunk. The people who succeeded in moving northwest through Europe did so by bringing two additional technological innovations to the region as well as their farming skills.

They had developed their own form of pottery, decorated with incised lines, so their culture is known to archaeologists as Linearbandkeramik or LBK. But a more important innovation was that the LBK people replaced the flint heads of their tranchet axes by stone tools with thick, heavy heads made of larger-grained metamorphic or igneous rocks such as jade, greenstone, basalt, or rhyolite. And rather than shape the heads by flaking them, they ground them down and polished them, making them smooth, but not particularly sharp. It is easy to see how heavier heads would have improved the performance of axes, but not why the LBK people went to such trouble to grind and polish them. Why work for hundreds of hours to produce an ax head that does not even look as if it would cut very well? One way of answering this question is to test modern replicas of stone axes. Pioneering experimental archaeology tests have indeed found that polished axes are much more effective and durable than tranchet axes. However, even with them, felling a tree was time-consuming. For instance, in the early 1950s, Svend Jørgensen and his colleagues showed that one person would take around eighty days to clear an acre of oak woodland, a quarter of the rate of someone using a modern steel ax. Polished axes had to be used in much the same way as tranchet axes, cutting obliquely through the trunk to produce a series of slivers of wood that gradually penetrated deeper into the trunk.

But useful as the experimental reconstructions of archaeologists are, they do not tell us why polished stone axes are able to cut through wood, or what the best shape should be. To investigate this further, therefore, I decided to investigate their design by combining a theoretical analysis of the process of splitting wood, with simplified cutting tests using metal wedges of varying shapes and surface properties. We investigated the force and energy needed to split coppiced hazel rods down the middle in my Instron Universal Testing Machine. The experiments were performed by my project student Joao Oliveira, who found some surprising results. Though thin, sharp wedges needed a lower initial force to start a cut (as one might predict), sharp wedges needed *more* energy to split the wood than thick, broad ones, and for a good reason. Most of the energy to split the wood was needed to overcome the friction between the wedge and the wood, and since thinner blades contacted the wood nearer to the tip of the crack, they had to push outward with a greater force to keep the crack moving forward, causing more friction. Thicker, broader blades needed less energy because they prized open the crack farther from its tip and needed a much lower force. The lack of a sharp blade on polished stone axes would not have been a problem because for most of the process the tip did not touch the wood at all. More predictably, smoother wedges needed less energy than rough ones to split the wood because they slid along the wood more easily and encountered less friction. Overall the results showed that the broad, smooth blades of Neolithic polished ax heads are highly efficient at splitting wood. We follow their design to this day when we cut up our firewood. Modern splitting mauls also have broad, heavy heads with a blade angle of around thirty-five degrees. One should never try to split wood using an ordinary steel ax; with its narrow blade it all too easily gets stuck in the log.

These results also have implications for the design of ax handles. After all, if a blunt ax head is good at splitting logs, then wedging the head into a hole in a wooden handle and hammering it against a tree trunk is all too likely to prize the two sides apart and split the handle. This has proved to be a constant problem for experimental archaeologists when they carried out their investigations. That it was also a problem for Neolithic people can be seen from the ax handle that was discovered in 1997 at Etton, Cambridgeshire; it had split lengthwise at the edge of the hole and had evidently been discarded by its user. But Neolithic people seem mostly to have been adept at making sure that this did not happen too often. They made the hole in the handle quite broad, so that the blade contacted it at the two ends, not at its sides, ensuring that ax blows did not produce a lateral force. And the hole itself was often reinforced to make sure that it did not split. The wood around the distal end of the handle was thicker around the hole, and flanges of wood extended above and below it. The Neolithic people also chose the type of wood carefully. They used wood such as oak with large rays and cut the hole tangentially through the wood, ensuring that the rays reinforced the top and bottom of the hole and prevented cracks from forming. In the axes found at the Neolithic lakeside villages around Lake Constance, the stone blade was rammed into an antler plug, which was in turn plugged into the handle. The antler must have acted as a shock absorber, reducing the impact stresses transmitted into the handle.

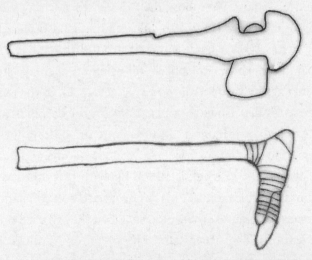

The design of Neolithic woodworking tools. In the Shulishader ax (top) the blade is held within a hole in the thicker distal end. In an adze (below) the blade is lashed to one side of a tree branch.

There are other ways of making a strong joint between the ax head and the handle. Native Americans bound the joint up, wrapping cord around the ax head and the end of the handle, making the familiar tomahawk, thus stabilizing the blade and preventing the handle from splitting. And later in Europe, people also developed a quite different type of ax. They carved a thick, heavy ax head and made a circular hole all the way through it to hold a wooden handle, using a wooden drill with sand abrasive. Recent experiments by Rengert Elburg and his group in northern Germany have shown that these "hammer axes" could be used more like a modern ax to cut down trees. The ax was swung horizontally, hitting the trunk at right angles and pulverizing the wood cells in front of it, so that a broad notch could gradually be cut.

But axes were just one of several woodworking tools that the LBK people developed. Perhaps the most common tool of all were adzes. Adze heads can be effectively mounted onto handles even more easily than ax heads by using the branching points of trees. This method exploits the way that trees strengthen their branch junctions and forks, a design that was only recently discovered, by my former PhD student, Duncan Slater, a lecturer in arboriculture (tree surgery) from Myerscough College, Preston, UK. Duncan showed that within the crotch of these joints, the wood fiber cells going to the two arms wrap around one another, preventing the joint from splitting. In Neolithic adzes, the head was strapped into a notch on the outside of the V, making a reliable joint. Rengert Elburg of the Archaeological Heritage Office, Saxony, Germany, and his colleagues have found that the LBK people made a variety of adzes in a range of sizes and mounted onto handles at different angles. Experimental tests showed that large adzes with the blade mounted at an acute angle could be used with an overhead action to fell trees, while ones with the handle at an obtuse angle worked better when used like medieval adzes; swung from above along the surface of a felled tree trunk, they could split wood off along the grain to hew the trunk into a square beam. Smaller, narrower-bladed adzes could carve these beams into a range of shapes or be used to gouge out wood from the inside of a trunk to make a hollow log boat. Finally, Neolithic people made chisels by tying short stone heads or beaver teeth to the ends of wooden handles, and by carving the long bones of oxen into a blade.

Polished stone tools are fairly efficient at cutting wood, especially along the grain. Using a typical Neolithic "tool kit," Phil Harding of Wessex Archaeology has been able to replicate the manufacture of the handle of the Shulishader ax, which had been found in 1982 on the Isle of Lewis, Scotland. He split the blank from a log of oak, shaved it to shape, cut the rabbets with a chisel, and pierced the mortise with a flint piercer, before finally sanding and polishing the handle. All this took some three or four days of work, but an experienced Neolithic carpenter could probably have done this faster. The LBK people were cer-

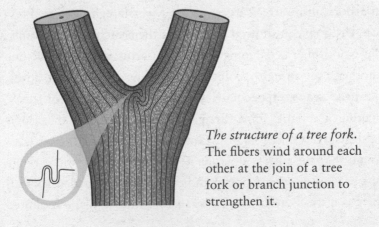

The structure of a tree fork. The fibers wind around each other at the join of a tree fork or branch junction to strengthen it.

tainly able to make far larger and more impressive artifacts and structures than ax handles. They made the earliest plows or ards by making a simple cross joint between two rods of wood and then shodding the end with a stone tip. Pulled by an ox wearing a wooden yoke, these simple scratch plows would have been able to cut drills into light soils for planting cereals.

The most impressive achievement of the LBK people must have been the wooden halls they built: the first houses of multiple occupation. No complete—or even partial—buildings have survived from the Neolithic period, but we do have an idea of their ground plan because of the survival in the soil of post-holes and the grooves left by wall planks, and this has enabled archaeologists to reconstruct them, for instance, at the Všestary Archeopark, Czech Republic. LBK longhouses could be up to fifty-five yards long and eleven yards wide. The roof was supported by three lines of posts within the building, while the outer walls were constructed using a line of thinner posts with grooves cut into them into which the builders slotted horizontal planks. These buildings are easy to reconstruct, as they left remains very like those of better-preserved Anglo-Saxon houses. They are also very like other wooden structures built more recently by people using similar Stone Age technology: the longhouses made by the Chinook in the Pacific Northwest of America and the Iroquois from the Great Lakes; the bamboo longhouses of Southeast Asia; the villages of Amazonian tribes; and the meetinghouses of the New Zealand Maori. The three lines of posts would have held up a pitched roof. The ridgepole set on top of the central posts and the purlins set on top of the outer posts supported rafters that joined to and overhung the outer walls. Finally the roof would have been filled in with wooden laths and covered by split wooden shingles, thatch, or turf. The structures would only have needed simple joints or rope fastenings between the posts to stand up and were probably built in a communal effort, like the buildings of the Iroquois and Maori, and decorated in the same way with wood carvings.

Thanks to a recent find we now also have a far better appreciation of the sophisticated woodworking skills of the LBK people. In 2011 Willy Tegel from the University of Freiburg, Germany, and his colleagues dug up the intact linings of seven wells from LBK sites in eastern Germany, where they had been preserved by the anoxic soils. The wells, all square in shape and about a yard wide, were made around 5000 BC from planks of oak that had been hewn flat by the use of small adzes and joined together with woodworking joints still used by modern carpenters. The base frame planks were firmly joined using mortise-and-tenon joints that were then secured by inserting wooden pins through the tenon. In the frames higher up the structure, the planks had notches cut halfway through them and were sequentially joined to the frame below by slotting the notches together, just like the pieces of those wooden dinosaur models you can buy from hobby shops (and the Lincoln Logs log cabin sets we used to play with as children). The whole structure reveals that even seven thousand years ago people were highly adept carpenters and able to use sophisticated techniques to build complex structures. The only messy aspect of their design was at the ends of the planks. Without efficient sharp tools, Neolithic people plainly found it hard to cut wood across the grain. Their blunt wedge-shaped blades could simply not make precise cuts. The ends of the planks therefore showed signs of having been charred and burned to weaken them enough to hack through them.

Later Neolithic houses that have been uncovered in Britain were smaller than LBK longhouses—usually consisting of just a single room, some twenty feet by sixteen feet, and probably only housing a single family—but recent excavations have shown that they were well fitted up with built-in furniture. We would never have known this had it not, paradoxically, been for the excavations of the stone-walled houses in the famous village of Skara Brae on Orkney, off the north coast of Scotland. Nowadays, Orkney is almost treeless, and even in Neolithic times before forests had been cleared, the trees would at best have been fairly short and stunted. The local people used another more readily available material to construct the walls of their houses—the old red sandstone that was laid down in river deltas in Devonian times and whose thin layers can readily be split to produce slender planklike slabs. In Skara Brae the houses were clustered together; each house, with its two-foot-thick stone walls, would presumably have been covered with a pitched roof supported by rare and valuable timbers. In the inhospitable Highlands and islands of Scotland, roof timbers continued to be critical for survival right up until modern times, so one of the cruelest but most efficient methods absentee landlords found to enforce the Highland clearances of the mid-nineteenth century was to burn the crofters' roofs. On the wall opposite the door of House 7 at Skara Brae was what looked like a sort of "dresser" with two rows of horizontal slabs supported by upright ones, rather like a modern CD rack or bookshelf. And along the left-hand wall were stone "box beds" like those still seen in modern crofts and in boats. It looked as if the stone was being used to mimic wooden planks, something that was confirmed when archaeologists excavated the remains of wooden houses at Durrington Walls near Stonehenge. In the earth, along with the postholes and wall slots, were the outlines of the remains of exactly the same furniture as at Skara Brae and located in exactly the same position. It seems that just like today, builders tended to make houses to identical plans.

The early Neolithic people built their houses using the trunks of mature trees, and heavy planks that they had split from their logs. But later, they also developed a new way of managing woodland, coppicing, to produce smaller, more manageable pieces of wood that they could use to build their houses more quickly and easily. If certain trees are cut down, they do not die. Instead, many broad-leaved species, including oak, ash, chestnut, hazel, and willow, and the conifer yew, resprout from dormant buds in their trunk. They throw up a number of fast-growing shoots that grow straight upward, lengthening and thickening rapidly. The shoots can be harvested repeatedly in crop rotations that can be designed to produce a supply of rods of a consistent diameter and length. Coppicing has several advantages over allowing trees to grow to maturity before killing them, grubbing out the stump, and planting new trees. First, since the coppice stools already have a root system to supply them with water, the shoots grow rapidly, even in their first year after the cutting. Second, since they do not have to transport water all the way up through their trunk, the new shoots are better supplied with water than the branches of a large tree and can grow faster, producing more wood per unit area of ground. This makes coppicing an ideal system for supplying firewood or, as we shall see in the next chapter, charcoal. Third, since the coppice shoots grow faster than branches, their leaves are farther apart. As my PhD student Seray Ozden showed, this allows the shoots to grow straighter, stiffer, and stronger than branch wood, giving them a wide variety of structural uses.

The best evidence of coppicing in the Neolithic is seen in the use of coppice rods in another first for humans—the first tracks and roads. In 1970 peat excavations near Glastonbury on the Somerset Levels, an area of drained marshlands in the west of England, uncovered a linear wooden structure. The Sweet Track, as it became known after its finder, Ray Sweet, proved to be a wooden walkway, designed as a causeway to help people travel between the lakeside villages in the region. The walkway was constructed from planks of oak sixteen inches wide and ten feet long that had been split from four-hundred-year-old trees and placed end to end. The planks were supported using crossed coppiced rods of oak, ash, and elm, the whole structure being pegged together with other thinner rods. Examination of the pattern of growth rings on the planks, a science called dendrochronology, dated the structure precisely to 3806 BC. Other tracks have been found throughout lowland England, the earliest being the Post Track, which was built in 3838 BC.

Neolithic people also used coppiced poles for a variety of other purposes, making the handles of tools, for instance. But the most generally useful technology that they developed using coppiced poles was wickerwork, a development of the methods people had used to weave branches together since before we came down from the trees. Wickerwork is more organized, though, and in its simplest form involves weaving a series of narrow shoots in and out at right angles through a parallel framework of thicker ones. On a large scale, hazel poles were used to make wicker panels or hurdles that were used for fences and gates that would, among other things, protect the regenerating coppices from people's oxen and pigs. Hurdles could also be used to make lightweight walls of houses, the panels wedged between the frames of half-timber houses and covered with mud or plaster to produce wattle and daub. People still make hurdles, and to do so they first insert a line of round poles known as zales into holes in a wooden proforma. They then wind thinner lengths of wood, usually poles that have been split into quarters using a billhook, back and forth through the zales, twisting the wood when it reaches the last pole and winding it back over in the other direction. When the wickerwork has been built up the entire length of the poles, it is finally freed from the proforma, and the hurdle can be lifted and pegged into the soil wherever it is needed. This basic weaving technique has proved to be even more useful when applied to the younger, thinner coppice shoots of trees such as willow. Neolithic people used them to develop a wide range of wickerwork, to produce, among other things, fish traps and baskets.

There is also evidence that Neolithic farmers made small lightweight rounded boats with wickerwork frames and a leather hull, very like modern day Irish coracles. The Greek traveler and historian Herodotus described such craft, and pottery model boats of this form have also been found. Neolithic people probably used these small boats for trade, as the blades of adzes and axes, which had been made from quarries next to the Danube, have been found along the banks of rivers throughout Europe. And the farmers themselves may well have used the boats for their long, slow migration across Europe between 6000 and 4000 BC. They would have punted them up to the tops of the tributaries of the Danube, carried them over the watersheds, and finally floated downstream in them along the rivers of northwest Europe. But of course it was not only Europeans who used wickerwork. It is a technology that is found all around the world, suggesting that like the ax and the adze it was independently invented many times and widely shared.

In summary, though advances in woodworking were not essential to the development of farming in the Near East, they *were* essential to its spread across Europe. Improved stone tools helped migrating farmers clear the land of trees, opening up a whole new way of life, and enabling them to migrate and colonize the continent of Europe in far greater numbers than the hunter-gatherers they supplanted. The story was much the same in other areas of the world, though farming emerged at different times in different places. The people also exploited different ranges of crop plants and developed rather different forms of farming with contrasting impacts on their regions' forests. In Japan, for instance, polished stone axes appeared as long as thirty thousand years ago but the first permanent settlements, farming millet, only appeared between 6000 and 4000 BC. In China farming appears to have started around 7000 BC, with people in the North growing millet and those in the South growing rice. In the Americas, maize, beans, and squashes were the main crops, but farming methods differed widely; some did not even involve clearing forests. Many of the tribes in New England grew their crops in forest gardens, while the Wintu and Cahuilla of California developed a "balanoculture," caring for oak forests and living on a diet made from acorn flour. In Africa the main crops were sorghum, millet, and yams, while in New Guinea they were sugar cane and banana.

But in every area it was the new polished stone tools that enabled people to clear forests and cultivate the land. They enabled the farmers to meet a huge variety of their needs, from building large houses, fencing their fields, making their tools, fashioning their furniture and houseware, building their boats, and even making their roads. The Neolithic world, though it would have had a lower forest cover than the Mesolithic, would have been even more dominated by people's need for and use of wood. And with its rectangular houses, wooden furniture, boats, and extensive road network, it would have looked surprisingly familiar to us.

CHAPTER 6

Melting and Smelting

If polished stone axes are the icons of the Neolithic, the periods that followed—known in Europe as the Copper and Bronze Ages—are typically represented by the weaponry that people could make using these new materials. Daggers are a particular favorite, along with spear points, shields, and helmets. One would think from this fixation that man's proper study was killing other men, and that the benefit of civilization was that people could now do so more quickly and efficiently. Quite apart from the fact that our ancestors could kill each other perfectly well using wooden and stone weapons—clubs, spears, and bows and arrows, for instance—another, far more important story is to be told about the benefits of copper and bronze. This chapter will outline how the people of the Near East and Europe came to be able to smelt and shape metals, and how they applied them for more peaceful purposes. As we shall see, not only did the people use wood to make these new materials, but they then largely used the new materials to improve how they harvested and shaped wood. Paradoxically, metals made people become even more reliant on wood and use far more of it. The new technology spread across Asia, and eventually into Africa; the result was to transform Old World civilizations and give them a decisive lead over those of the New World.

However, we might never have been able to smelt metals had it not been for our relationship with a quite different set of materials: ceramics. Even before people settled down into farming communities, they must have noticed the potential benefits of clay, a common soil material that is most obvious when it lines the banks of rivers and lakes. People would have seen how easy clay is to shape when it is wet, yet how hard it sets when it dries out. Right across the prehistoric world, people found ways to use it. Clay could be slapped over wicker walls to make them draftproof, or it could be molded into bricks that could be reinforced with straw and left to dry in the sun. Houses with sun-dried brick walls were common in the Fertile Crescent throughout biblical times, where they made up for the scarcity of wood in the region, and houses built of adobe are still used throughout the arid parts of the world. Clay can even be used successfully in wetter regions if it is protected from rain. Houses made of cob—basically bits of stone and mud molded together—are common in rainy Devon, southwest England, where they are topped off with ludicrously picturesque overhanging thatched roofs that keep off the soft Devon drizzle. I'll always remember my surprise when, on a visit to a cream tea shop in a picture-postcard village, I even spotted a garden wall that had its very own thatched roof!

There is a better way of waterproofing clay bricks, though: heating them up. Clay minerals consist of plates of mica that in their natural state are bound together with relatively weak hydrogen bonds; these are strengthened or weakened by removing or adding water. However, if the clay is heated to over 900°F, all the water bound up in the structure is driven off, and permanent bonds form between the clay particles. The clay is converted to a biscuitlike solid—earthenware—that is unaffected by water, but is still somewhat porous and weak. It has to be heated to over 1,800°F for some of the chemicals to fuse or vitrify to form a glasslike material that joins the clay particles together and makes a new material—stoneware—that is stronger and impermeable to water.

People learned how to make fired ceramics relatively early in prehistory; the first-known clay sculpture is the Venus of Dolní Věstonice, a four-inch-high female figurine found in Moravia, Czech Republic, which dates from the late Paleolithic, some thirty thousand years ago. The first clay pots did not appear until later, though, the first fragments appearing in East Asia between twenty thousand and ten thousand years ago. Elsewhere, clay pots did not appear in any numbers until the Neolithic, since they are heavy and fragile and so manifestly unsuitable for mobile hunter-gatherers. But when people settled down, the benefits of clay pots became more obvious. Being waterproof, they could be used to store both dry food and liquids, and they could also be put onto fires to cook food. A new range of porridges, stews, and broths no doubt greatly expanded the culinary range of Neolithic chefs. Fired bricks and tiles, which were made with clays containing a greater proportion of sand, seem to have come later. The first fired bricks were made about 4300 BC for the walled settlement of Chengtoushan, China, while the first roof tiles replaced thatch in Mesopotamia only in the third millennium BC.

There is just one problem with making ceramics—it is extremely difficult and dangerous to heat them up to high enough temperatures to waterproof and strengthen them. As we saw in chapter 2, most wood fires operate—at least in their early stages—at temperatures between around 400°F and 600°F, and only later, when the volatiles have evaporated off and only carbon is left, do temperatures rise to a maximum of around 1,100°F. Neolithic people fired their pots in holes in the ground—the first kilns—which enabled them to raise temperatures to around 1,500°F. However, to obtain consistently high temperatures they were also the first to convert wood into a new, more concentrated, source of energy that consists of pure carbon—charcoal.

The key to making charcoal is to heat wood to temperatures above 600°F, by which point all the volatile compounds have been forced off, but below 900°F, when the carbon that is left would start to burn. People have always achieved this in essentially the same way: by limiting the air supply to a wood fire. In the simplest charcoal kilns, colliers pile logs closely together and cover them with turf before setting fire to the bottom of the pile. The whole burn can take several days, during which the overseeing colliers peek into the pile and increase or decrease ventilation to keep the burn at the optimal temperature. Charcoal burning is long and dirty, reducing the mass of wood by over 60 percent and wasting over half the chemical energy it stores, but it produces lumps of pure carbon with twice the energy density of dried wood. Charcoal also retains the open cellular structure of wood, which maximizes its surface area and allows oxygen to power rapid combustion. Using charcoal rather than wood allowed people to heat kilns to over 1,800°F, even without extra ventilation, enabling them to produce stronger, more waterproof pottery. Charcoal-fired kilns also enabled craftsmen in Mesopotamia to be the first to produce another material that would later become almost as useful: glass. As early as 2300 BC they learned that heating sand with ash made by burning wood or seaweed would fuse them into a hard, shiny material, one they could later remelt to decorate and waterproof ceramics, or to mold into jewelry or waterproof containers.

But by far the most important use people made of charcoal, at least in the Old World, was to smelt metals. Because charcoal is made of the highly reactive element carbon, if they used it to burn metal oxides, it did not merely heat them up, but also removed the oxygen from the ore to produce pure metal. The first metals people used were not made in this way, though. They found native copper in the mountain regions of Anatolia and shaped it by heating it up and beating it with stone hammers. Demand for this new metal grew rapidly because copper combines some of the best material properties of both stone and wood, properties that we now know are related to its atomic structure. Because metals such as copper are composed of crystals of identical atoms, they are almost as stiff and strong as stone, but they have an added advantage. Their structure is not flawless, and there are faults and dislocations within the crystalline matrix. When the material is stressed, these faults can run away through the material, relieving the local stress and absorbing large amounts of energy. The consequence is that metals can be beaten into shape, a property known as malleability, and stretched into wires, a process known as ductility. Most important of all, because metals absorbs so much energy, they are even tougher than wood and are strong in all directions, not just along the grain. This means they can be shaped into all sorts of useful tools and resist bending forces without snapping—properties that make them ideal for a wide range of purposes. In particular early craftsmen found that metals could be shaped into long slender cutting tools with sharp cutting edges that could also readily be resharpened by scraping them with stones. The only disadvantage of metals is their high density, several times that of wood, but even so, most metals are still, weight for weight, just as stiff and strong as wood and even tougher.

As demand for copper grew, people discovered that if certain stones were heated in charcoal fires, pure copper was released, in what must have seemed a miraculous way. By the early fifth millennium BC, copper ore was being mined in Bulgaria and Serbia, and soon after, metallurgists realized that they could make metal tools by pouring the molten metal they had smelted into ceramic molds that they had shaped and fired earlier— once again using charcoal—to harden them and make them heat resistant. Two processes, both powered by this single wood derivative, enabled metallurgists to cast objects from a whole new supermaterial. By around 5000 BC a new age had dawned, at least in the Near East and Eastern Europe, the Copper Age or Chalcolithic.

But good as copper is as a material, it has the disadvantage that dislocations run through it just too easily, so though it is tough, it is also fairly soft. The consequence is that copper blades blunt and snag all too easily. Over the next two thousand years, therefore, metallurgists developed ways of combining copper with other metals to produce new, harder alloys. One element that was used in Iran from as early as the fifth millennium BC was arsenic. Just 0.5–2.0 percent arsenic increased the stiffness of the copper by 15–30 percent and enabled it to be work hardened more effectively, making better cutting tools. However, arsenic is toxic, and the alloying process was hard to control. From the middle of the fourth millennium BC, therefore, a more effective alloy was developed, bronze, in which 12 percent tin was added to the copper. This produces a tough, corrosion-resistant metal, which quickly became the material of choice for making tools and weapons. The only problem was sourcing the tin; it is a much rarer element than copper, and the two metals are rarely found in the same places. Early European metallurgists would have had to set up a supply chain to transport tin from England, Germany, or Spain to Eastern Europe, where it could be combined with the copper.

People quickly took advantage of the properties of copper and bronze to improve and add to their inventory of tools, particularly those they needed to cut down trees and shape wood. By far the most common early metal tools were ax heads. The ice man Ötzi, for instance, who was discovered in a glacier in South Tyrol, Austria, in 1991, and had died around 3300 BC, not only had a flint knife in his wooden rucksack but a copper ax. The metal blade was narrower than stone blades of the period, which had necessitated a redesign of the handle. Since the blade was too thin to plug into a hole in a straight handle, as stone ax heads are, it was instead attached to a handle made from a branch junction, like in a Neolithic adze. The rear of the blade was fitted into a slot at the end of the branch and tied in with leather straps. The same system was used in the early Bronze Age to mount ax heads, or palstaves as they are known, onto their handles, but the slot was an obvious weak point; later on, therefore, bronze axes were made quite differently; the head was cast hollow so that it fitted over the branch like a glove. Still later Bronze Age ax and adze heads were made with a socket at their base to mount on a straight wooden handle, more like a modern ax.

Whichever way their heads were mounted, the new metal

axes proved far superior to the old polished stone equivalents. James Mathieu of the University of Pennsylvania has shown that bronze axes cut through tree trunks around twice as fast as stone ones, and just as quickly as iron axes, and they can be used with a similar technique to that used by modern lumberjacks. Bronze axes can be swung much more horizontally at tree trunks, and their more slender blades can cut a much narrower kerf, an angle of on average seventy degrees rather than the eighty-five degrees for stone axes, wasting something like 25 percent less wood.

People soon added bronze adzes to their woodworking tool kit, as these were even better than stone ones at shaving and shaping wood along the grain. However, a much more important addition were bronze chisels. Since bronze can resist tensile forces, bronze chisels could be made much more slender than their polished stone equivalent, and since they were also tougher, they could stand up far better to being hammered. They could therefore make deeper, sharper, and more accurate cuts across the grain of wood. At last people could make precise joints such as the mortise and tenon, overlapping joints, and dovetails. The huge benefits this conferred can be seen in that the appearance of copper and bronze tools coincided with the emergence of two wooden technologies that were to transform transport in the Old World and kick-start the emergence of international trade: plank ships and wheels.

We saw in the last chapter that Neolithic log boats were perfectly good at transporting people and goods short distances up and down rivers and across lakes. However, their round hulls made them inherently unstable, and as they were limited in size by the diameter of tree trunks, they were narrow and low in the water, so that they could never be seaworthy. The solution—to attach planks to the boat to increase its width or depth, or both—was obvious, but it is extremely difficult to make watertight joints between planks using blunt stone tools. With sharp bronze tools this would no longer have been a problem, so it is no surprise that the earliest plank boats and ships that have been found date from the Bronze Age. Perhaps the most obvious way of making log boats more stable is to split the hull down its center and add one or more planks between the two sides. This seems to have been the technique followed by shipbuilders in Northern Europe, and many Bronze Age boats have started to be discovered preserved in the mud surrounding the shore, particularly in Great Britain.

Among the best-known, and oldest, Bronze Age boats, are the three Ferriby Boats, which were discovered in the 1930s and 1940s by two schoolboys, Ted and Will Wright, on the northern shore of the Humber estuary, Yorkshire, just a few miles from my home. The planks of these boats, the earliest of which dates to 2000 BC, were stitched together using twisted yew twigs, but the precision aspect of the design was the joints between the planks. The oak planks were closely held together by overlapping tongues and grooves, and the joints were stabilized by rods that were wedged into parallel sets of cleats that had been carved out of the thick planking. The shipwrights had finally added curved internal frames to stabilize the shape of the boat. The Ferriby Boats, which were almost fifty feet long and with a width of almost six feet, could probably carry cargoes of up to 3.3 tons. A flat-bottomed boat, the Brigg Raft, was also found in 1888, on the other side of the Humber estuary, showing the diversity of boat design even three thousand years ago. It was plainly used to ferry livestock over the river Ancholme to and from the town of Brigg, where a horse fair is held to this day.

These craft probably moved goods around the inland waterways of Britain. It is unlikely that any of them would have been able to trade long distance across the English Channel and around the Mediterranean, yet seagoing vessels must have been used in these waters during the Bronze Age. They would have been needed to transport English tin, which was mined around the county of Cornwall in the southwest, over to centers of the bronze smelting industry such as Cyprus, which is named after the metal copper. One candidate for just such a seaworthy vessel is the Dover Boat, which was found in 1987; its length is uncertain as it was not complete, but it was certainly much wider than the Ferriby Boats.

Few early Bronze Age ships have been found in the Mediterranean, which is not surprising, since the conditions there are hardly suitable for preservation. We know there must have been ships, though, because in 1975 the pioneer of underwater archaeology, Peter Throckmorton, discovered pottery of the Cycladic type at the bottom of the Aegean off the coast of Hydra. Though the ship had long since rotted away, this had to be the remains of a shipwreck dating to around 2200 BC. Fortunately, another plank ship has survived from the Bronze Age to give us some idea of the sophistication of the woodworking in this period. The funeral ship of Khufu, which dates from 2500 BC, was found complete, though disassembled, in a pit next to the Great Pyramid of Giza in 1954 by the archaeologist Kamal el-Mallakh. Egypt could never have been the center of the shipbuilding world as it has few trees, and its ships were needed mostly just to transport goods up and down the Nile. Yet even here and at this early date, Khufu's ship showed sophisticated design and manufacture. It was built from large numbers of short planks made from cedar of Lebanon, all numbered and ready to be joined together with precise mortise-and-tenon joints. The ship was painstakingly reassembled by a team led by the Egyptian Department of Antiquities' chief restorer, Ahmed Youssef Moustafa. It was 143 feet long and 19.5 feet wide, and though plainly a ceremonial rather than working ship (it was presumably built to take the resurrected king across the heavens), it shows just what Bronze Age shipwrights must have been capable of.

The methods the Egyptians used were similar to those used in the Mediterranean a thousand years later. In 1982 Turkish archaeologists discovered a late Bronze Age vessel, the Uluburun Shipwreck, which dates from around 1400 BC just off southwest Turkey. The forty-nine-foot-long hull was built in the same materials and with the same mortise-and-tenon joints as the Khufu ship. It was carrying a cargo of copper ingots, which it was probably transporting from Cyprus to Mycenaean Greece, a good indication of the trade networks that plank ships had opened up. More than any other technology, it was plank ships that enabled the Mediterranean to become the crucible of civilization in the West. They could carry people and goods quickly and freely across a vast maritime region, accelerating material and intellectual progress and allowing large cities to be supplied. The Roman Empire would have been politically unsustainable without huge ships to transport wheat from its Egyptian colony to supply the citizens of Rome with a free bread supply. And plank ships later did a similar job around Arabia, India, and the Far East, maintaining the communications and trade links that were vital to coordinate growing empires.

● ● ●

If bronze tools would have made building plank ships far easier, they would have been essential for constructing the structure that transformed land transport and went on to make machinery practical: the wheel. The concept of moving objects with wheels no doubt developed from people observing how fruit and other circular objects can easily roll along the ground. It is often suggested that the intermediate stage between balls and wheels was to use log rollers to move stones. But there have recently been a whole raft of other suggestions about how ancient people could have moved heavy stones—by rolling them. Dick Parry has suggested that the Egyptians could have rolled the limestone blocks used to build the pyramids to the construction site by strapping curved wooden cradles around all four sides. Meanwhile, Isle of Man–based engineer Garry Lavin has suggested a similar technique—though in this case using wicker cages—could have been used to roll the famous bluestones of Stonehenge from their quarry in the Preseli Hills of Pembrokeshire to the site itself in Wiltshire, two hundred miles to the east. And Andrew Young of the University of Exeter, UK, has suggested instead that the stones could have been moved on the tops of stone spheres that rolled along grooves in wooden rails. The consensus among archaeologists, though, is that large stones were hauled across land using sleds, as is often depicted in Egyptian tombs, and that the architects used various ways to reduce friction. The Egyptians are said to have poured water onto the sand to make a slippery surface, while Neolithic Britons could have dragged their stones over wooden rails lubricated with fat.

Wheels operate in a way that is somewhere in between rolling and sliding. In a wheel, unlike a sled, the rim does not have to slide across the ground, but the wheel does slide around its axle and this motion is resisted by friction. Wheels reduce the force resisting motion because the friction operates so much closer to the center of rotation of the wheel, and because the axle can more easily be lubricated than the ground. The larger the diameter of the wheel and the thinner, smoother, and better lubricated the axle, the more easily the wheel will run. An axle made of wood has to be reasonably thick to withstand bending forces, around an inch or two in diameter; to make an efficient wheeled vehicle, early wheelwrights would have had to construct wheels that were well over twenty inches in diameter. You might think that this was easy (as indeed did my editor!). All you would need to do was to cut disks of wood out of a tree trunk, like slices from a salami. Unfortunately, in the Neolithic and even in the Bronze Age, this would simply not have been possible. Without saws, wooden disks could never have been sliced from a log because there were no tools that could cut so large a piece of wood straight across the grain. Even if they could have been cut, wheels made from slices of tree trunk would never have worked. They would have been extremely weak and would quickly have split right down their center. And they would have split even if they had never been loaded. Wood shrinks as it dries out, but not evenly in all directions. Since most of the cellulose microfibrils are oriented along the grain, wood shrinks far less in this direction than across the grain, by around 0.1 percent compared with 4–8 percent. And because wood rays are also reinforced by cellulose fibers, wood shrinks only half as much radially as tangentially. Consequently when a wooden disk dries out, it splits radially and a wedge of around fifteen degrees opens up. A wheel made out of a slice of the trunk would not be able to roll at all. In fact, all large beams made from freshly cut wood split as they dry out, which is why the wooden pillars of barns and the roof beams in old houses always have splits in them—though fortunately in those cases the flaws don't materially weaken the structure.

The first wheels were therefore cut from planks of wood that had been split radially from a tree trunk and then chiseled into shape. These wheels would have worked, but since they would have been limited in size to half the diameter of a tree trunk, they would have had to be rather small. Most Bronze Age wheels were therefore made by joining together two or more, usually three, planks of wood. The difficulty with this design is making the joints strong enough to stop the wheel from folding up. Bronze Age wheelwrights overcame this problem by cutting large rectangular mortises right across the surface of the wheel, then inserting tight-fitting battens into them, which were then pegged into place. The wheel would still be weaker across the join, but at least it would be usable.

All of these processes involve precise woodworking, so it is no surprise that the first evidence we have of wheels in the archaeological record dates back to 3500 BC, long after copper tools had appeared, and just as bronze ones were starting to become common. Wheels appeared almost simultaneously in the Sumerian civilization of Mesopotamia, in the Caucasus, and in Eastern Europe. Evidence from written texts suggests that the first vehicles were four-wheeled wagons, whose design had been converted from that of sleds; the symbols for the two vehicles in Proto-Indo-European writing are much the same, apart from circles at each corner in the symbol for a wagon. Indeed the earliest evidence of wheeled vehicles is just such a diagram that was incised into the Bronocice Pot, a clay vessel dated to around 3400 BC that was found in southern Poland. The earliest actual wheel found to date is the Ljubljana Marshes Wheel, which dates from around 3150 BC. This twenty-eight-inch disk was made from two planks of ash and was attached with wedges to an axle of oak. In this case, therefore, the wheels rotated with the axle, which spun within grooves at the bottom of the vehicle, which was a two-wheeled pushcart. Later in the Bronze Age, people made wagons and carts with stationary axles; they had to attach large hubs to the ends of the axles to allow the wheels to rotate independently and freely without falling off. This design was more complex and involved more precise woodworking skills, but because the wheels on the two sides could rotate at different speeds, it made it far easier to steer vehicles around corners.

As wheels became more common, roads themselves necessarily followed, especially in wetter regions where the ground was softer. The earliest paved roads were found in Ur in Mesopotamia and date from around 4000 BC, while in northwest Europe, Bronze Age people moved on from the narrow wooden trackways of the Neolithic to produce wider corduroy trackways. These first appeared around 3000 BC in Germany and Holland and were made simply by laying rows of half logs, each about twelve to thirteen feet long on the bare earth round side down. A later trackway at Cloonbony, Ireland, which dates from around 2550 BC, is more complex, having rails almost five feet apart, rather like a modern train track, which were pegged at intervals to the ground.

Later in the book we will examine how the designs of ships and wheeled vehicles developed, but it is worth emphasizing here the crucial role that metal tools had in their invention—and the best way of doing that is to contrast the technology of the Old World, where metallurgy developed, and the New World, where it did not. When the Spanish conquistadores invaded Central and South America in the sixteenth century, they discovered a continent with highly sophisticated civilizations, but ones that had never exploited hard metals such as bronze or iron. The lack of hard metals had not stopped the Incas, Aztecs, and Mayans from building magnificent cities with huge stone pyramids, crafting beautiful pottery, and making exquisite gold jewelry.

But despite their many advances, none of the New World civilizations had developed plank ships. The Incas, for instance, relied on reed rafts to transport huge stones across Lake Titicaca, while the Aztecs carved flat-bottomed log canoes up to fifty feet long to transport people and goods around their complex system of canals. The Mayans used similar craft to carry out their trade around the Gulf of Mexico. Until recently it was thought that plank boats had never been built in the Americas before the European conquest. However, we now know that there are two exceptions. The Chumash Indians of the Santa Barbara Channel area of California did make simple sewn plank boats, tomols, which were up to twenty-five feet long and four feet wide. Though these take up to six months to build using just stone and shell tools, they were far more seaworthy than the canoes of their neighboring tribes. Similar plank boats known as dalcas were also built by the inhabitants of the Chonos Archipelago, on the coast of Chile, some 47° south. They were built by sewing together three planks of larch and were rather like the early Bronze Age boats from Great Britain.

How could these isolated tribes build plank boats, and how did they learn how to build such advanced craft? Some archaeologists think that their traditions might have been imported from contact with Polynesians, who had also developed the technology to make simple plank hulls. The Polynesians benefited from making the blades of their adzes out of the thick part of the shells of the giant clam *Tridacna gigas*. Mollusk shells have been found to be much tougher than stone because they incorporate a range of techniques to divert cracks and increase the work of fracture. This means that thin shell blades can withstand bending forces and can perform almost as well as metal ones. Several lines of evidence back up the idea that the techniques used to build plank boats were passed on to the coastal American tribes by the Polynesians. For a start, these parts of the Americas are closest to the most easterly Polynesian islands—Hawaii and Easter Island. The boatbuilding tradition of the Chumash also dates to around thirteen hundred years ago, the time when the Polynesians first reached Hawaii. There certainly seems to have been contact as the Polynesians were raising sweet potatoes, which they must have obtained from the American continent, some thousand years ago, while the Chumash started to use complex fishhooks like those used by the Polynesians at around the same time. There are even some indications from linguistics, as the word *tomol* does not resemble any other Chumash word but does appear to be related to the Hawaiian word for "useful tree." It is intriguing to think that wooden technology can give us such clues about the colonization of the world.

It is much-better known that none of the great American civilizations used wheels, either for transport or to make pottery. They carried their goods in packs or used beasts of burden such as llamas to do it, and the Incas pulled large stones using sleds, like the ancient Egyptians. But in the Americas they never made wheeled vehicles as people did in the Old World. This was not because the New World civilizations had not invented wheels. Several of them, including the Aztecs, made children's toys that ran on clay wheels, but they never made them in wood and mounted them on sleds. Most commentators ascribe the omission to two reasons: first, that the landscape was simply too hilly; second, that the people had no draft animals to pull carts or wagons. Neither of these reasons is convincing. After all, the hilly country didn't stop the European settlers of the continent from finding wheels useful after they had conquered it, while some parts of Central America—Mexico City and the Yucatán Peninsula, for instance—are notoriously flat. But to address the more serious of the two suggestions, it should be noted that many Bronze Age vehicles from Europe, including the earliest ones found, were handcarts, while in China the wheelbarrow was invented around AD 100 to 200. Wheeled vehicles are useful even if you have to pull them yourself. And draft animals were available; in North America, the Plains Indians used their dogs to drag their travoises, simple sleds made from the poles of their tepees, along the ground. It seems much more likely that it was instead the technical difficulties in constructing workable wooden wheels using just stone tools that prevented these civilizations from developing wheeled transport.

The lesson of this chapter is therefore clear. As is so often the case, new technologies do not replace old ones, but inspire new ways to use them. In the case of copper and bronze, the major impact of the new materials was to enable people in the Old World to exploit their main structural material, wood, more effectively, allowing them to revolutionize their transport networks. The result was to give the people of the Old World a massive lead in logistics, one that five thousand years later was to help them discover the New World and subdue its people.

CHAPTER 7

Carving Our Communities

You only have to visit an open-air museum to see how wood dominated the lives of our forebears. These museums, which aim to re-create the lives of our forebears, can be found all around the world, and all of them have the same basic design. They are composed of old buildings that have been relocated to the museum site, where they are strategically placed to reconstruct cottages, farms, workshops, hamlets, and even whole villages. And to do this more realistically, they are fitted out to look as though they are still in use—with furniture, tools, utensils, and ornaments in place, and even with roaring fires in the grates. I love them all and have visited sites across the world, ranging from my first, the Frilandsmuseet, in the outskirts of Copenhagen, which I was taken to when I was just eight years old, to my most recent, St Fagan's National Museum of History, outside Cardiff, which I drove to myself when I was fifty-six. I've been to relocated villages in the Canadian Rockies, and in Kota Kinabalu, in Borneo. And I've visited museums that have reconstructed buildings from archaeological evidence, museums such as Butser Ancient Farm, in West Sussex, and, my personal favorite, both for its charm, seclusion, and unlikely delights, West Stow Anglo-Saxon Village in Suffolk, England. I've never managed to get to an open-air museum in the United States, but my editor tells me that Williamsburg, Virginia, is well worth a visit.

Apart from the sheer joy of visiting the museums, which brilliantly satisfy our love of learning, our nostalgia, our enjoyment of a day out in the fresh air, and our sheer nosiness, they give two strong impressions even to the casual visitor. First, they show just how little the life of people changed from the start of the Iron Age, around three thousand years ago, to the industrial revolution, around two hundred years ago. They show that in many rural areas, life was much the same as it had always been within living memory; and many parts of the world still seem stuck in the preindustrial past. Second, they show just how much the lives of ordinary countryfolk depended on wood. Their houses were made of wood or at least had wooden frames and were roofed with wooden shingles. Their furniture was almost entirely wooden—the beds, tables, chairs, and cupboards—as was their kitchenware—the barrels, jugs, cups, bowls, and spoons. The fuel storages outside their houses were full of the split logs that they burned to keep themselves warm and cook their food. On their farms, the vehicles—the carts and wagons—were all wooden, as were the handles of their tools—their plows, hay rakes, mattocks, and scythes. And their power plants—the water mills and windmills—were overwhelmingly wooden constructions. Even the few items that were not made of wood had all been manufactured using it. The iron cutting tools and pots and pans had all been smelted using charcoal; the cloth had been spun on wooden spinning wheels and woven on wooden looms; and the leather had been tanned with tree bark.

Yet in many ways wood is an unpromising material with which to make complex three-dimensional items. Unlike clay or metal it cannot be molded into shape; complex wooden items either have to be assembled by joining together several smaller pieces or carved out of a single large piece. And because wood is anisotropic—it is far weaker and more brittle across the grain than along it—it is tricky to carve and vulnerable to splitting. It is no surprise, therefore, that the wooden tools we have looked at so far in this book—long, thin structures such as spears, digging sticks, bows and arrows, and log boats— were made from largely unmodified branches or tree trunks and were designed to resist bending forces just like the trees from which they were made. It is a testament to our ancestors' ingenuity, and to the merits of a new metal—iron—that they managed to carve out a whole new wooden world to give them warm, comfortable lives.

People first used iron around fifty-five hundred years ago, when they hammered rare lumps of meteoric iron into shape to produce beads and other precious items. An iron dagger was among the objects in Tutankhamen's tomb, for instance. However, iron was only smelted for the first time—like bronze in charcoal-fired kilns—around 1500 BC. It proved to be more difficult to make useful tools from this new metal than from copper or bronze because its melting point was much higher— over 2,200°F—so it could not be melted in charcoal fires and cast into molds; it had to be heated to as high a temperature as was then possible, around 2,000°F, at which point it was soft enough to be hammered into shape. However, as blacksmiths perfected their craft, two advantages showed themselves. First, the new metal had better mechanical properties than bronze, especially after it had been worked, so it could be made into finer and harder-wearing cutting tools. Just by chance, by beating the iron in the forge and folding it over, smiths incorporated fibers of slag into the iron, fibers that reinforced the metal just as cellulose fibers reinforce the walls of wood cells. The result was a material—bar iron—that was more resistant to corrosion than the pure metal, and much stiffer and tougher.

The second advantage was that iron ore is far commoner in the earth's crust than the ores of copper and tin, so it can be mined and smelted locally. Smiths could therefore make iron tools far more cheaply than bronze ones, so iron technology spread rapidly. Beginning with the discovery of iron smelting in the Middle East, by around 1000 BC it had spread right across Europe, reached China by around 700 BC, and sub-Saharan Africa between AD 200 and AD 1000.

The first iron tools were simply iron versions of the bronze woodworking tools that craftsmen had already developed—axes, adzes, and chisels—and woodworkers used them in just the same way. They split the wood, shaped it, and carved joints in it when the trees were newly felled and the wood was still green and soft. Some trades continued to use these ancient techniques long after new ones had been developed. Indeed, the craft of green woodworking continues to this day and has long proved to be ideally suited for building large structures such as bridges, houses, and ships.

For some purposes, people found that wood was best used unmodified, in the form of whole logs. We saw in the prologue that ships' masts were essentially complete tree trunks. Apart from the convenience of this arrangement, which meant that they needed only to have their branches cut off and be stripped of bark before they were mounted into a ship, there were good mechanical reasons for this; the trunks of trees are actually prestressed to help them stand up to the wind. One minor disadvantage of wood is that because it is made up of hollow cells, it is weaker in compression than in tension. When they are squashed, the cells crumple at loads only around a third of their tensile strength. Trees overcome this in their trunk by prestressing the outside in tension. After the outer layers of wood are laid down as the tree grows, they try to shrink but are prevented from doing so because they are attached to the layers of wood inside them. The outer cells are therefore held in pretension and tend to compress the interior cells. This results in a pattern of longitudinal prestress in which the inside of the trunk is held in compression and the outside in tension. This arrangement strengthens the tree in a storm. On the windward side, the bending of the trunk puts the trunk under even higher tensile stresses, but the cells can easily cope with this. The advantage to the tree is that on the leeward side, the tensile prestress *reduces* the compression that the cells experience. As a result a tree can bend almost twice as far in the wind as it would if it had not been prestressed and resist almost twice the wind load. So masts and booms made from tree trunks were capable of withstanding all but the worst storms.

The only problem with prestressing occurs when we cut trees down. As the ax or saw cuts the bottom of the trunk loose, it is freed to respond to the prestress. The outer parts of the trunk contract longitudinally, and the inner trunk expands, and this can cause the lower ends of the trunk to bend outward, splitting the trunk along its length to form what foresters call shakes. Large species of *Eucalyptus* trees are particularly prone to this problem. Not only did this make Australian gum trees useless

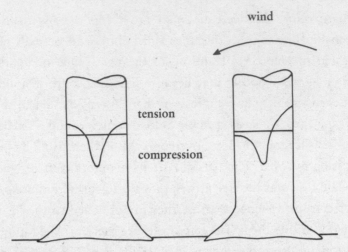

Prestressing in the trunks of trees puts the outside of the trunk into pretension and the center into compression (left). This reduces compression in the leeward side of the trunk when the wind subjects the tree to bending (right).

for making masts, much to the consternation of the British colonists, but as the trunks can spring outward by as much as one to two yards, it can make felling them lethal.

Solid logs can also be used to make the frames of houses and the structural elements of bridges. Most wooden bridges were short, unsophisticated structures that used logs as simple beams to resist bending forces. However, in the Song dynasty, the Chinese developed a unique form of woven-timber arch bridge that proved capable of spanning gaps of almost two hundred feet. These rainbow bridges, as they are known, were composed of two sets of polygonal arch members, each of which is made up of logs joined end to end with simple mortise-and-tenon joints, and which weave through each other. Logs are tied between the two arch members to keep the structure taut, and a wooden pathway is then constructed over the top, while a roof keeps the structure dry, just as in the covered bridges of New England. According to historical records, the first timber arch bridge was built by Xia Shouqing, military commander of Qingzhou, Shandong Province, between 1032 and 1033, and a rainbow bridge is depicted in a painting of a bridge over the Bia River in Bianliang City, the capital of the Song dynasty, which is dated to around AD 1120. No bridges from this period survive, but the tradition continued in country areas, and just over one hundred of these beautiful structures remain, crossing deep river gorges in Fujian and Zhejiang Provinces, eastern China.

Oak-framed houses also continued to be built in the green woodworking tradition, though as time went on the builders increasingly used timbers that had been hewn into square beams. We saw in chapter 5 how Neolithic woodworkers built their rectangular longhouses using lines of posts driven into the ground. These in turn supported the outer walls and a central ridgepole. The rafters were then simply supported between the ridgepole and walls, and their bending strength held up the roof. Builders in continental Europe continued to use this method, but in Britain, Bronze and Iron Age farmers reverted to building circular houses, not unlike those built in the Mesolithic. Archaeologists have long been puzzled by this and have usually just described it as a peculiar British fashion. However, it seems more likely that it was actually a consequence of a relative lack of wood in Britain. In a country with a high population density and where the forest cover was already falling to around 25 percent, it made sense for farmers to fill in the outer walls of their houses with wattle and daub, and to make them circular to minimize the perimeter. The circular shape also allowed the builders to support the conical roof without needing a long central pole to support it at the center. They simply bound the rafters together with horizontal rings of coppice poles, which prevented the rafters from pushing outward on the walls and stopped the center of the roof from sagging. The poles acted in just the same way that the cloth in umbrellas helps them maintain their dome shape.

Housebuilders were conservative, just as they are today, but over hundreds of years they did introduce two major improvements to their designs. First, they started to realize why their buildings lasted only twenty to thirty years; the alternate wetting and drying of the base of the posts produced ideal conditions for the growth of fungi, and the tops of their buildings' foundations simply rotted away. Soon monumental architects were placing the wooden posts of their temples on top of stone pedestals; now their buildings, protected from rain by their roof and from wet soil by their foundations, were capable of surviving for hundreds of years.

The second great advance in wooden architecture was the development of the roof truss. The posts supporting the ridgepole in rectangular Neolithic huts cluttered up the living space, but if they were omitted, the weight of the roof would cause it to sag, and the rafters would push the walls outward. The solution—probably first devised by the Romans—was to tie the lower ends of opposing rafters together using a horizontal tie beam running across the building. Additional struts could also be added within this A-frame arrangement to provide further support to the rafters, to prevent them from sagging in the middle. Triangular roof trusses were a common feature of the Romans' basilicas, their public buildings, and of the early churches whose design was modeled on them. However, following the fall of the Roman Empire, the advanced techniques the Romans used seem to have been lost; Anglo-Saxon houses were narrow, for instance, and the roof was supported in the middle by posts that were buried in the ground, just like Neolithic houses; consequently they rotted away within a few decades in just the same way. Fortunately, the church builders of the Middle Ages reintroduced both stone foundations and roof trusses to Northern Europe, no doubt copying surviving Roman basilicas and churches. The result was a medieval flowering of wooden architecture, both sacred and secular. Many early churches were built in wood, such as the Saxon church at Greensted, Essex, England. The most celebrated wooden churches, though, must be the great Scandinavian stave churches, such as the earliest, that of Urnes in Sogn in the western fjords of Norway, which was built in 1130. The churches rise magnificently beneath a bewildering profusion of steeply sloping wood-tiled roofs, but their basic design is simple. They have a central nave held up at the corners by vertical posts or staves, which are held together by crossbeams at the top, while the roof is supported by steeply sloping wooden rafters that make up the angled arms of an A. The outer aisles (there may be two sets) are simple lean-to additions. The simple construction is often hidden, though, by Saint Andrew's crosses and panels between the central posts, which make it look as if the walls are supported by arches, as they actually were in the stone churches that the craftsmen were mimicking.

The same simple structural design can be seen perhaps to even greater effect, and certainly more clarity, in the great tithe barns of southern England. The fifteenth-century Harmondsworth Great Barn, the largest timber-framed building in Britain, stands just a few hundred yards from the runways and terminal buildings of Heathrow Airport and is well worth a detour from the crass consumerism of the departure lounges.

This huge barn is almost two hundred feet long, thirty-seven feet wide, and thirty-nine feet high, having eleven bays with complex trusses supporting the roof over the central nave and side aisles. It could hold almost four thousand tons of grain. The English colonists brought this basic design to North America, where it gradually evolved into the all-wooden three-bay barns and New England barns, with their chestnut roof shingles.

Traditionally, the houses and halls of Northern Europe were also built with wooden frames, but on a smaller scale. In the simplest, cruck houses, the walls and roof are supported by pairs of timbers that were split from trees that had gown with a natural curve. The timbers were placed in lines opposite each other, the tips touching to form a rough A shape. The rest of the structure could then be built around them. The more sophisticated frame houses, what are commonly known as half-timber, were built with trusses supporting the roof just like a tithe barn, but instead of using single long timbers running from floor to roof, they were constructed as a series of boxlike frames, one on top of the other to make a large multistory house. The genius of the design was not only that shorter, cheaper beams could be used, but that the upper floors could jut outward a few feet from the lower ones. This not only gave extra floor space, but the weight of the upper story helped push up the center of the floor beams, preventing the floors from sagging at the center. There was just one problem: the upper stories of town houses on either side of the street came closer together, so fires could jump much more easily between buildings—a feature that no doubt contributed to the rapid spread of the Great Fire of London in 1666.

Despite fire-safety fears, though, there is nothing intrinsically wrong with the design of wooden buildings and a lot going for them, and green oak houses continue to be built and to be extremely popular with wealthy clients in country districts to this day. Oak is easily worked in the green state, hewn into beams using adzes that give an attractive irregularity to the surface, and the houses are an excellent example of prefabricated construction. The joints can be carved and tried out off-site, enabling the structure to be rapidly assembled once the foundations are in place. Another advantage of using green wood is that after the joints are set with wooden pins, the wood shrinks as it dries out, causing the joints to tighten and strengthen the structure. Finally, once erected, the frame can be filled in with highly insulating walls of wattle and daub that make the house cozier than any structure of brick or stone.

But perhaps the greatest triumph of the green woodworking tradition was the Viking longship, which transported hordes of murderous Norsemen on their voyages of conquest and discovery. These fine vessels enabled Vikings to subdue much of Britain, Ireland, and Northern Europe, colonize Iceland and Greenland, and discover the New World centuries before Columbus. And Viking traders spread their influence as far east as Russia, and as far into the Mediterranean as Constantinople. Yet the Vikings built their sleek craft using only the most old-fashioned of tools: axes to cut down the trees, broad-axes to shape the wood, and augers to drill holes in it. These activities are clearly depicted in the part of the Bayeux Tapestry that shows William the Conqueror building his invasion fleet; the Normans, who invaded England in 1066, were, after all, descended from Vikings who had settled in northern France at the beginning of the tenth century.

Having chosen and felled suitable straight-grained trees, the Viking shipwrights first carved and laid down the keel of the ship, a length of elm or oak with a T-shaped cross section. Next, they sequentially attached side strakes to it. These were made from oak planks, which were radially split from a tree trunk, a feature that gave them several advantages over modern sawn planks. Since the timbers exactly followed the course of the fibers and vessels, no end grain was exposed, so the plank was totally watertight and also extremely strong and flexible. The large rays of the oak also reinforced the planks laterally and prevented them from splitting. To shape the strakes the Vikings used a new iron tool, the drawknife. Drawknives work exactly like adzes but are made quite differently, having a long iron blade that is sharpened along one edge, with wooden handles attached at each end. To shape the wood along the grain the drawknife is held with the blade at a large angle to the wood so that it digs in and is then drawn toward the woodworker, enabling a long sliver of wood to be removed. The earliest draw-knives recorded were uncovered in the Mästermyr chest, which dates from around AD 1000 and was found on the island of Gotland, Sweden, but they had probably been used for centuries before that. The Vikings also used modified drawknives—molding irons—to carve a groove or rove along the lower inside edge of each strake. They then jammed wedges of rope caulking into the groove to provide a watertight seal between it and the next strake, then pinned each strake to the outer surface of the one inside using iron rivets, to form a strong "clinker" hull. Finally, the shipwright selected bent branches to make ribs for the boat, and straight crossbeams, which they attached at each end to the upper strakes along the sides of the hull to stiffen the structure. In the magnificently carved Oseberg ship, which dates from around AD 800, the mast was attached to the hull and supported via an ingenious "mast fish," a T-shaped piece of wood that was carved from the branching point of an oak tree, utilizing the strengthening mechanisms used by the tree itself in life. But perhaps the height of the Viking age is shown by the Gokstad ship from AD 890, which is less ornate but which has purer, more graceful lines and a more effectively stiffened structure.

The great age of longship construction was over by around AD 1100. Partly this was because the Vikings had settled down to a Christian life, reducing their desire for rape and pillage, but also because they were finding it increasingly difficult to find timber with straight enough grain to split. Later medieval shipwrights had to revert to using trees with irregular grain, which they sawed up to produce planks of inferior quality. The new ships had to be built using a solid frame. However, shipwrights continued to make extensive use of curved pieces of wood, cut from great oak trees that had deliberately been grown in the open. These provided the bent knees, futtocks, and breast boards, among other timbers that supported the heavy frames of men-of-war well into the nineteenth century.

Traditional green woodworking techniques proved less suitable for producing smaller structures—pieces of furniture, plates and bowls, ornaments, and the like. For a start they create a lot of waste. For instance, to create a regular plank from radially split timber, much of the wood has to be shaved off, and cutting the wood across the grain using axes leaves a wide kerf. Green woodworking techniques also tend to be rather imprecise, and there is the additional difficulty that the wood tends to shrink and warp after it has been shaped, which can loosen the joints. Craftsmen therefore gradually developed a new set of woodworking techniques that shaped wood after it had been dried out or "seasoned" rather than before—carpentry. And to do so they invented and perfected a new set of tools that fully exploited the superior stiffness and hardness of iron and steel.

The first item to be added to the carpenter's toolbox was the crosscut saw, the first tool that could cut wood efficiently across the grain leaving only a narrow kerf. Metal saws resembling the simple bone saws used by Mesolithic people first appeared in ancient Egypt. They were made of bronze and had teeth arranged so that the saw cut through the wood as it was pulled toward the carpenter. This arrangement, which persists in Japan to this day, prevented the narrow blade from buckling but limited the power that the carpenter could apply. The teeth of these saws were also quite simple, so they acted rather like serrated knives, concentrating the pressure in small areas to improve the depth of cut and helping sweep the sawdust away. The design of the teeth was improved, however, once saw blades started to be made of iron. In a modern crosscut saw the front of each tooth is sharpened to form a blade; each tooth is effectively a tiny chisel. The teeth are sharpened alternately to the right and left of the saw so that they sever small lengths of cell wall, which can then be swept free in the form of sawdust. Craftsmen added two other features to their saws to maximize their efficiency: the teeth are bent or "set" slightly outward so they cut a kerf that is slightly wider than the blade of the saw and help it sweep through without getting stuck; some crosscut saws also have rakes between each pair of teeth to help sweep the dust free. The Romans overcame the difficulty in making a blade that is narrow enough to cut through wood, yet rigid enough to resist buckling when it was pushed through a cut, in two ways. In backsaws, which are still common today in the form of tenon saws, the blade was held rigid by a thick top edge and provided with a wooden handle. In frame saws, a narrow, flexible blade was held taut by a rigid wooden frame, a design that is nowadays seen in bow saws, hacksaws, and jigsaws. The disadvantage of both types, the rigid frame limits the maximum depth of the cut, was overcome in some frame saws by arranging the frame so that it came out on either side of the blade, not above it. Only later, in the eighteenth century, with the advent of spring steel, could saws be made rigid enough, without a back or frame to support them, to produce typical modern saws.

Crosscut saws were particularly useful in felling trees and cutting planks of wood precisely to length, but increasingly carpenters also needed to saw wood along the grain. This would help them cut accurate dovetail joints to join planks together at right angles, and it would also help them cut precious woods into thin veneers that could cover cheaper timber. Most important, it would enable foresters to cut planks from trunks and branches that had less than perfect grain, and which could not therefore be readily split—timber that was becoming increasingly common as old-growth forests were felled and woodland was put under management. To cut wood along the grain to produce the sort of planks we still use today, a new design of saw was needed: the ripsaw. Whereas in crosscut saws it is the side of each tooth that acts as a chisel, in a ripsaw it is the bottom of the tooth, which points forward, that is sharpened. It cuts across each fiber cell and removes the loosened region at the same time. Ripsaws became more and more common from medieval times onward, when long two-man "pit saws" were used to cut large tree trunks into boards, the lower man working in a pit below the horizontal trunk and the top man standing on top of it. The process increasingly became mechanized, especially in continental Europe, where the first water-powered sawmill using reciprocating ripsaws was set up by the Dutchman Cornelis Corneliszoon in 1594. Needless to say, apart from the iron blades, sawmills were built almost entirely of wood.

To cut planks more precisely down to size, to smooth them, and to carve them into more complex shapes, the Romans added another tool: the plane. Planes cut wood in exactly the same way as adzes and drawknives, splitting it along the grain and removing the upper shaving. The advantage of planes, though, is that the frame holds the narrow steel blade at the optimal angle of around forty degrees and controls the depth of wood that it removes. Like drawknives, planes can be designed with a range of blade shapes, so that, as well as smoothing boards flat, they can cut curves into wood and even cut precise tongues and grooves along the edges of planks, ideal to swiftly form overlapping joints to link them from side to side.

Armed with just these few simple tools, carpenters could make all manner of objects in two and three dimensions simply by joining together planks of wood, arranged in such a way that in each of them the grain runs along the long axis, ensuring maximum strength. Doors are an interesting case in point. You might think that it would be simple to make a door. All you would need to do would be to cut a plank from a large tree trunk. However, such planks would be hard to obtain and extremely expensive, and they would be prone to warping unless they were cut radially from the trunk, and vulnerable to splitting down their length. Carpenters instead devised two main ways of making cheap, strong doors out of smaller planks of wood. The most obvious way was to join multiple planks of wood side to side using tongue-and-groove joints, and to strengthen the door laterally by riveting horizontal planks called battens to it, to make what is known as a board batten door. Large metal hinges prevented the door from dropping, and a diagonal cross member called a brace could also be added to do the same thing. These doors work well but are heavy and use a lot of wood, and though still common in country cottages, board batten doors have largely been replaced by panel doors. These have a frame, composed of vertical stiles on either edge, horizontal rails that are joined to them by mortise-and-tenon or mitered joints, and mullions that run vertically between the rails. Carpenters then fitted panels of thin board (or glass) into grooves in the frame. The whole structure can also be decorated by carving the inside edges of the frame with specialized planes known as molding irons. Panel doors are lighter, more graceful, and less prone to warp than board batten doors, though they can drop in the same way if the joints work loose.

Carpenters also quickly learned how to make three-dimensional structures by joining planks and battens to each other with mortise-and-tenon, dovetail, miter, and a host of more complex joints. But the demands of jointing mean that it is far easier to make joints at right angles than at any other angle. Like houses, therefore, wooden furniture—chairs, tables, dressers, desks, cupboards, bookshelves—are almost invariably rectangular or cuboidal. Furniture made in this way can be extremely elegant—think of the delightful chairs made by the eighteenth-century Shakers or by the art nouveau architect and designer Charles Rennie Mackintosh with their ladder backs. And conventional rectangular furniture certainly fits well into rectangular houses, though as we shall see later in the book, this design has disadvantages. To make items with a rounded shape, or ones composed of curved pieces of wood, craftsmen had to develop rather different techniques.

Small objects with a circular cross section—plates, bowls, cups, chair legs, and woodwind instruments—were made using a different technique from carpentry: turning. The idea of shaping timber by using special chisels to remove slivers from a rotating block of wood until it became circular was no doubt inspired by the way that potters shape their clay vessels. The potter's wheel had been developed at around the same time as cart and wagon wheels, around 3500 BC, but only around 1500 BC did the first lathes appear and were represented in Egyptian wall paintings. The arrival of iron no doubt helped in the development of lathes as it could be made into chisels that were hard enough to shape the wood without wearing away too fast.

All lathes have the same basic design. The piece to be worked is clamped at one end onto an axle and pinned at the other so that it is free to rotate. The turner spins the piece and presses a hard-pointed chisel into it to shave wood off evenly around the axis of rotation. The differences between lathes are principally in the way the rotation is powered and when the wood is carved. In the earliest Egyptian lathes the work piece was rotated using a strap wrapped around it, which an assistant pulled back and forth in a reciprocating motion; the turner carved the piece during only one of the strokes. In a bow lathe, the turner himself spun the piece using the string of a bow wrapped around it, just as drill bits are turned in bow drills. This obviated the need for an assistant but left the turner with only one hand free to hold the chisel. The big advance was the pole lathe, in which the turner powered the lathe with a treadle; pushing down on the treadle pulled down on a cord that was wrapped around the piece of wood or billet being turned, while the other end reached up to the end of a long springy pole and bent it downward. Releasing the treadle allowed the pole to straighten and the piece to spin backward. Regular injections of force could therefore set up a reciprocating action in the wood, while leaving both hands free for working the chisel. Pole lathes, manned by craftsmen known as bodgers, were in use as far back as the Iron Age in Europe and were common in the Viking period; the street Coppergate in York, England, where the Jorvik Viking Centre is located, was named after the large number of cup makers who plied their trade there.

Wooden cups were made from pieces of wood placed with the grain parallel to the long axis of the lathe, so that their narrow stems could be as strong as possible. In contrast, wooden bowls and plates were made from blanks with the grain at a right angle to the long axis of the lathe to strengthen them across. For centuries country people ate and drank almost exclusively from wooden dishes; each person would have his or her own plate or bowl, which was used for all meals. Lathes were also used to make handles for cutlery, and, especially as people replaced their wooden dishes with ceramic crockery in the eighteenth century, to turn legs for tables and chairs.

The other technique developed by craftsmen to introduce curves into their designs was steam bending. Heating wood to the boiling point of water loosens the bonds between the hemicellulose molecules, softening the matrix, and leaving the cellulose fibrils free to shear past each other. Lengths of wood can therefore be bent by placing them in a steam bath, removing them, and bending and clamping them into the desired shape. As the wood dries and cools, it will retain its new form.

Steaming wood has been used to make a wide range of curved objects, from the ribs of boats to snowshoes and tennis rackets. It is a technique that has enabled craftsmen to produce furniture that is more graceful and comfortable than traditional designs, the most famous design being the classical Windsor chair, with its graceful bow back. Early Windsor chairs from the seventeenth century were made in the beech woods of High Wycombe, in the Chiltern Hills, west of London. Carpenters shaped the bow backs using the pale beech timber from the area and assembled them with the stretchers and legs made by local bodgers, and with seats made by local benchmen, and shipped them via Windsor to London. The design was then introduced to the American colonies, the first chairs being shipped in by Patrick Gordon, lieutenant governor of Pennsylvania, in 1726, after which local production soon started up in Philadelphia.

But the most important items to be made using the technique of steam bending were barrels. In classical times, wine, oil, and other liquids were held and transported in earthenware amphorae—tall, round-bottomed jars with two handles around the neck—but they were heavy, hard to move, impossible to stack, and highly breakable. Bits of broken amphorae make up the great majority of archaeological material from the time. Wooden barrels, made by joining together curved wooden staves, were probably invented around 350 BC by the Celts and quickly proved far more practical. They were strong, could be rolled along the ground to move them, and could readily be stacked. The key to their success is the curvature of the staves, which allows the center to bow outward, enabling the longitudinal stiffness of the wood to resist the internal pressure of the liquid inside. To make a barrel, coopers had first to precisely carve each stave to shape, using specially made drawknives to give it the desired curved cross section, and a jointer to cut the sides at an angle so when put side to side the staves formed a circle. They then joined the staves together, inserting them into a temporary hoop at each end, and heated the staves until they fitted together snugly. Finally, the cooper made the ends and fitted them onto the ends of the staves and reinforced the structure at the ends with iron rings. Barrels proved to be the lifeblood of commerce in preindustrial times, the equivalent of the tin cans, plastic bottles, and shipping containers of today combined.

Wheelwrights used all three of the woodworking techniques: carpentry, turning, and steaming, to improve the design of a final wooden structure: the wheel. As we saw in the last chapter, early Bronze Age wheels were made by joining together three planks of wood and were both heavy and weak. The designers of ancient chariots were keen to improve on this design to enable them to produce lighter, faster, and more robust vehicles. By the late Bronze Age, they had come up with just such a design, the spoked wheel, in which the hub, which was turned from a split-resistant wood such as elm, was joined to long thin spokes that were shaped using drawknives or lathes. In turn the spokes were mortised into a rim, which could be made from thin branches that had been bent into a curve using steam. The first spoked wheels were made around 1500 BC by the Egyptians and their Middle Eastern rivals, but wheel design reached a peak in the chariots of the Homeric Greeks. Their wheels had only four spokes, which allowed the rims to flex slightly as they turned, giving the Greek chariots some suspension, smoothing the ride as they raced across the bumpy battlefields of Troy. The downside was that even a hero such as Achilles had to remove the wheels of his chariots overnight or turn the whole chariot upside down to prevent the wheels from being bent out of shape as the wood gradually deformed under the weight of the chariot. Later wheels produced by Roman and medieval wheelwrights were more solidly built and were reinforced with iron tires. Some of their wheels were dished, or saucer-shaped, with the spokes arranged in a slight cone to stiffen the wheels laterally and prevent the joints from coming loose as the carts rolled over rutted roads. And farmers often left their carts standing in shallow water to keep the wood wet and swollen and keep the joints taut, which is why the hay wain in John Constable's famous painting is standing in the river Stour.

Over the three thousand years following the discovery of iron smelting, therefore, iron tools helped to carve a world where wooden artifacts were ubiquitous. But as we have seen, all of the techniques were time-consuming and demanded high levels of mechanical dexterity and expertise. Consequently woodworking employed huge numbers of skilled craftsmen, as can be seen by the sheer number of common surnames arising from wood-based trades: as well as Carpenters and Joiners, there were Wrights, Wheelwrights, Shipwrights, Wainwrights (who made carts and wagons), Bodgers, Bowyers and Fletchers (who made arrows), Turners, Bowlers, Coopers, Sawyers, Foresters and Colliers (who made charcoal). Masons used wood-handled tools, Millers lived and worked in water mills and windmills made of wood, and Glaziers, Potters, and Smiths of all kinds used charcoal fires to heat their furnaces. The Iron Age was a world in which wood dominated people's lives, something that remained the case right up until the last two hundred years.

CHAPTER 8

Supplying
Life's Luxuries

There could not be a greater contrast between the open-air museums we looked at in the last chapter and the major cultural museums of the world. Rather than presenting the artifacts of everyday life, the big museums exhibit artworks, ornaments, and texts produced for and about the powerful, the wealthy, and the political elites. Judging from my visits to the British Museum, what the powerful seem to have mostly been interested in, at least until the last thousand years, were battles and hunts. In other words they loved depicting men killing other men, men killing women, men killing other animals, or in the case of the statues from the Parthenon—supposedly a high point of civilization—men killing imaginary organisms such as centaurs. Even well into the Renaissance there was a high breastplate quotient: depictions of men in armor that was indented with improbable pectoral muscles and six-packs. Fortunately many later works of art depict peaceful subjects: handsome men and beautiful women, both with and without clothes; animals and plants; landscapes; and abstract patterns. Whole rooms are filled with jewelry; gold and silver; ceramic and glass vessels; parchment books; and stone and bronze sculptures.

Amid the sensory overload, brought on by the sheer number of precious and finely wrought objects, few people notice the absence of the material that was commonest in everyday life: wood. It's always harder to spot when something is missing. For despite its manifold advantages, wood seems to have rapidly fallen out of favor with ruling elites. There are some practical reasons for this: wood is less shiny than metals or jewels, less transparent than glass, and less durable than stone or bronze. But it is hard not to suspect that the main reason was that wood was just too ordinary, too plain, too common. If even the poor could afford wooden artifacts, the rich would not want to own them! As we shall see, only for objects in which the mechanical superiority of wood was an overwhelming factor in the choice of material did they accept it. Even then, the elite ensured that the artifacts they purchased were made out of particularly choice and rare types of wood and decorated in such a way as to disguise its humble provenance.

So far in this book I have tended to talk about wood as if it were a single uniform material. But it is important to examine now how wood from different species of trees varies in characteristics such as density, stiffness, and color, and to examine the reasons for these differences. Wood is naturally adapted to the needs of the tree it supports, to its size and ecology, and to the soil conditions and the climate in which it is growing. The large broad-leaved canopy trees of temperate regions, such as oak, ash, beech, and lime, need to transport water relatively quickly up their trunks to supply their leaves, and they have to be able to withstand high winds. To do so, they produce wood with relatively large water-conducting vessels and fairly hollow fiber cells to produce a thick trunk as cheaply as possible. This gives the wood a medium specific density of around 0.5. Understory trees such as holly, dogwood, and box are shorter and need less water and tend to be slower growing and longer lived. They therefore lay down wood with much narrower vessels and thicker-walled fiber cells, producing denser, harder timber. Finally, faster-growing pioneer trees that colonize open ground, species such as birches, and riverside species such as poplar and willow, have particularly wide vessels and thin-walled fibers, allowing them to support rapid growth and make a thick trunk as rapidly as possible. This gives them soft wood with a specific density of as low as 0.35. The pattern is the same in northern boreal forests, where the canopy trees are conifers such as pines, spruce, and fir, and the pioneer trees are birches, maples, and aspen. In tropical rain forests, the wood shows greater extremes of density. It can range from below 0.1 in the fast-growing pioneer trees such as balsa, to over 1.0 in slow-growing understory trees such as ebony and ironwood; these woods actually sink in water!

Wood also varies in color, principally because trees differ in the amounts of colored defense chemicals—tannins, phenolics, and the like—that they pump into their heartwood to kill fungal diseases and prevent rotting. The longer lived a tree and the warmer the climate in which it grows, the more defense chemicals it needs and hence the darker the wood. Of the temperate trees the long-lived oaks and cedars have the darkest and most durable timber, while the fast-growing poplars and willows have the lightest and most flimsy. Tropical trees in general have darker wood than temperate species; consequently woods such as teak are widely used to make garden furniture. The timber of understory trees such as ebony and blackwood is particularly dark, in sharp contrast to white wood of the pioneer balsa tree.

The green woodworkers and local carpenters we introduced in the last chapter mostly used medium-density wood from large canopy trees. They used timber from the longest-living species such as oak and cedar for buildings, ships, and carts that had to stand up to the wet weather, and wood from shorter-lived trees such as ash and beech to make tools and furniture that were kept indoors. In contrast, the craftsmen who made furniture for the elite tended to use denser, darker understory trees to make, or at least clothe, their furniture and decorative artifacts. The best-preserved early furniture was that made by the Egyptians and stored in their tombs. Few trees grew along the narrow Nile corridor, but despite the lack of local wood, the Egyptians were wealthy enough to import a range of timbers from their neighbors and had access to many types of exotic woods from North Africa, the Middle East, and the Mediterranean. They made good use of local *Acacia* and maple, but they were the first carpenters to cover cheap wood with veneers: thin sheets of wood that they sawed off whole logs of more exotic timbers. They also developed the art of marquetry, gluing small pieces of wood of contrasting colors to form patterns and pictures. The tomb of Tutankhamen was filled with wooden furniture, including bed frames painted to mimic the bodies of lions, and beautiful boxes decorated with inlays of wood, glass, and jewels. But the most gorgeous item must be the throne of Tutankhamen, which today resides in the Cairo Museum. The throne is basically just an ordinary armchair, but it is heavily gilded and its backrest has an exquisitely tender picture of the boy king being anointed on the shoulder by his wife and (this being Egypt) his half sister Ankhesenamun.

The Romans continued to develop the art of marquetry, and it has long been a favorite form of decoration in China and

India, which have good supplies of colored timbers of all sorts. However, in medieval Europe, which had little access to tropical timbers, carpenters tended to make their furniture from solid local wood, using one of the densest and darkest timbers at their disposal, oak. Instead of colored patterns, they decorated it by covering it with a vast array of complex relief sculptures. Oak furniture, fireplaces, and paneling in rich homes became more and more elaborate, but the most sophisticated woodwork was used to decorate places of worship. Most churches had an oak rood screen separating the nave and the chancel, beautifully carved with Gothic arches, and with the panels in between containing paintings and sculptures of the saints. Above, the rood loft was decorated with a filigree of leaves and flowers. The decoration was most complex and beautiful in cathedrals, where not only the rood screen, but also the whole of the choir stalls, were constructed in beautifully carved oak and crowned with the heads of saints, clergy, and even mythical beasts. The craftsmen seem to have been unable to curb their love of ornamentation. Even the underside of the folding seats—the misericords—were decorated with depictions of animals, plants, people, mythical creatures, and even scenes of everyday life. The sheer energy of these carvings inspires the imagination. The author Lewis Carroll was brought up in Ripon, Yorkshire, where his father was a canon in the cathedral. It is said (admittedly mostly by the locals) that he got his inspiration for the opening scenes of *Alice in Wonderland* from a misericord in the cathedral—it depicts a rabbit fleeing down a hole, pursued by a griffin.

Oak carvings can be impressive and are certainly durable, but in many ways oak is a poor wood for sculpture. Rings of wide vessels are laid down each spring, and these form lines of weakness where the wood can split, and the huge rays form a visible "figure" that can distract the eye from the form of the piece. The most successful and detailed wooden sculptures are made from woods with more consistent, finer grain, so craftsmen have a long tradition of carving in light woods, whether the ruling elite valued the results or not. One of the standout objects of the Cairo Museum must be the sycamore statue of Ka-Aper, a scribe and priest from the early Fifth Dynasty, around 2500 BC. He is represented in an amazingly lifelike way, portly and bald, but he looks friendly and you could almost imagine meeting him walking down the street. This impression is supported by the fact that the Egyptian diggers gave him the name Sheikh el-Beled, "headman of the village," apparently struck by his resemblance to their own village elder. It is perhaps telling that this sculpture is of someone who was not of particularly high rank; statues of pharaohs were carved from stone and were highly stylized and impersonal.

The most successful and detailed wood sculptures of the early Renaissance were carved in limewood or linden, a timber that is softer than oak, more consistent, and lacks a strong figure. The heights of carving in this tradition were reached in the work of two fifteenth-century German sculptors, Tilman Riemenschneider and Veit Stoss. Their elaborately carved altarpieces, some of which can still be seen in churches in Germany and Poland, are comparable in humanity and ambition to the better-known stone masterpieces of the Italian Renaissance. They are particularly successful in bringing to life the suffering of Christ and his followers. Lime continued to be the timber of choice for wood carvers throughout Northern Europe, and woodcarving reached its peak in virtuosity in the work of Grinling Gibbons, a Dutchman who moved to England in the middle of the seventeenth century. Benefiting from using the finest of the new steel tools that were being produced in Sheffield, his carvings, most notably at Hampton Court Palace, London, and Petworth House, Sussex, are breathtaking in their lightness and perfection. Within formal friezes decorating fireplaces and paneling, Gibbons was able to depict details as fine as the grains within an ear of corn, individual petals in flowers, and even the strings of violins.

From the Renaissance onward, European furniture makers also turned their back on oak and chose more closely grained and beautifully colored woods, such as chestnut and walnut. And in homage to the classical world, they started to decorate their simpler designs with veneers and inlays of darker tropical woods and paler ivory, which they were now able to obtain from trading partners in the Far East and from their colonies in the New World. Decoration reached its zenith in the seventeenth and early eighteenth centuries, for instance, in the heavy baroque furniture of Louis XIV of France and the lighter, more graceful rococo style favored during the reign of Louis XV. With the classical revival of the late eighteenth century, cabinetmakers started to make simpler furniture that depended for its effect more on the quality of the wood and the perfection of its proportions. Firms such as Thomas Chippendale in London emphasized the beauty of a new tropical timber—mahogany. Exploiting the huge trees that grew in the hills of the British colony of Jamaica, they were able to saw large sheets of veneer that could cover the surfaces of their tables, while they used smaller pieces of timber to carve the legs and frame. They used inlay far more sparingly, largely to emphasize the lines of their creations and make truly elegant furniture. The Chippendales prided themselves on being able to furnish whole country houses and even made furniture for the servants' quarters. This was basically similar in design to their fashionable pieces, but was constructed more solidly out of Baltic spruce or pine and had less elaborately concealed joints.

Of all the arts, only in the field of music did the superior mechanical properties of wood make it the inevitable choice. Most musical instruments include a resonance chamber, in which the vibrations of a mass of air can be magnified at certain desired frequencies, and virtually all of them are designed to enable the walls of the chamber or a separate plate to vibrate in sympathy and further improve the volume and quality of the sound. To help achieve this, musical instruments have always been made of wood, which is light and stiff enough to conduct sound at high speeds and so resonate at high frequencies. The earliest musical instruments must have been naturally occurring sheets of wood that make up the buttresses of tropical rain forest trees. They are commonly struck by chimpanzees to make threatening noises during their competition bouts and were until recently also used by Bornean tribes to communicate over long distances. To achieve better sound quality, musical instruments have always been made from timbers that have a fine, even grain and so transmit vibrations efficiently; unlike furniture and tools, musical instruments are never made of ring-porous tree species such as oak or ash, which contain lines of large vessels along their growth rings that tend to soak up sound. Instead they use diffuse porous woods such as maple, box, or ebony, or softwoods such as spruce.

The manufacture of wooden musical instruments was perfected in the seventeenth and eighteenth centuries in the baroque period, as instrumental music first started to challenge the primacy of the human voice. Makers of blown instruments such as the Hotteterre family from France and the English maker Thomas Stanesby produced recorders, flutes, oboes, and chalumeaux (early clarinets), which were turned on sophisticated lathes and made from increasingly choice woods. The maple and sycamore used in Renaissance instruments such as crumhorns, shawms, and Renaissance recorders were replaced by darker, harder woods such as box, cherry, and blackwood. These denser woods conduct sound at higher velocities and so preferentially amplified higher frequencies, helping to extend the range of the instruments upward and give them a brighter tone. Baroque recorders, for instance, have a range of over two octaves, whereas in the Renaissance instrument it is only a twelfth. But as a recorder player, I have always preferred boxwood instruments to recorders made from harder blackwood or ebony; to my mind at least they give a mellower, more "woody" tone.

The early seventeenth century also saw the heyday of instruments that combined a "brass" mouthpiece with a wooden body. Cornetts or cornettos and their smaller relatives cornettinos were made from curved lengths of boxwood or hornbeam that were split down the center, hollowed out, and finally glued back together and covered in leather. The instruments had finger holes drilled in them and were fingered rather like flutes, though the sound was made with an acornlike mouthpiece. Cornetts were famed for their tone, which was said to closely mimic the human voice. They were commonly used in the antiphonal music composed for St. Mark's in Venice, in which different groups of singers and instrumentalists called to and answered each other across the huge church. They were vital to the success of the sacred music of the Gabrielis, of Monteverdi's *Vespers of 1610*, and of early operas.

But it was in string instruments that craftsmanship in wood

reached its peak. Keyboard instruments such as harpsichords and pianos have wooden soundboards that lie just beneath (or in the case of upright pianos, just behind) the strings. These are made by gluing together side-to-side planks of the softwood spruce, which have been quartersawn. These sorts of planks are the sawn equivalent of radially split planks; the cuts are orientated within the trunk so that the rays are aligned parallel to the width of the planks. This prevents the planks from warping and also speeds up the vibrations across the board, making the sound brighter.

The most sophisticated of all stringed instruments acoustically must be the plucked and bowed instruments such as violins, cellos, viols, lutes, and guitars. All have an upper surface, the soundboard that is made of quartersawn spruce, which is set moving by the strings via the bridge. To improve the tone, the soundboard is usually reinforced by gluing struts to its back, which makes it stiffer, and violins have an additional peg that links the soundboard to the belly of the instrument, further amplifying the sound. Holes in the soundboard help emit the sound to the audience. Despite extensive research no one has found the secret of the beautiful tone of Stradivarius violins. However, investigations of the wood have revealed that the soundboard is made from particularly slow-growing and fine-grained spruce that was harvested from the Alps. The fine grain of the wood and the bright tone of these instruments may be related to the cold weather and poor growing conditions of the Little Ice Age of the sixteenth to eighteenth centuries during which the trees were growing. With global warming, which speeds up tree growth, it may never again be possible to construct such perfect instruments.

But the fashions in music were changing. As the composers of the classical period such as Haydn and Mozart sought to wring more expression from their music by virtue of rapid changes in key, and as early Romantic composers such as Beethoven strove to increase the dynamic range of the music, instrument makers had to modify the lightly built baroque instruments. Violins, even Stradivarius instruments, were altered by raising the bridge, lowering the neck, and replacing gut strings with metal-wound ones to produce a bigger sound. The wood frames of early fortepianos were replaced with iron frames, which could withstand higher string tension, be fitted with heavier metal strings, and play more loudly. And woodwind instruments were fitted with large numbers of keys to make them easier to play and give a more even tone across all twelve semitones of the musical scale. So cluttered did they become with keys that some woodwind instruments such as flutes were even redesigned in metal. And the change saw the demise of cornetts and their bass equivalent, the serpent, replaced with brass instruments equipped with metal valves rather than finger holes.

So, despite their love of music, the rich and powerful lived in a world that was far less dominated by wood than the poor. But paradoxically the life they led would have demanded a far higher consumption of the material. Large quantities of wood, and of the charcoal that is derived from it, would have been needed to smelt and work the metal artifacts they used, to make their pottery, especially the fine porcelain that demanded several firings to finish, and to make glass. Like the proverbial dog in a manger, the rich did not want to use common woods themselves, but by using it to produce other more costly materials, they would have deprived the poor from using it. By surrounding themselves with beauty and luxury they would have left the poor colder and with less shelter.

CHAPTER 9

Supporting
Our Pretensions

O n the north bank of the river Tweed in the Scottish Borders, overlooking Norham Castle on the English side of
the river, is a strange church. Ladykirk, or Our Lady Kirk of
Steill to give its full name, looks odd from the outside as it is
strangely squat and roofed with huge flagstones. The inside,
meanwhile, is dark, its thick stone walls and low stone vaulted
roof being pierced by just a few small windows. The legend is
that the church was built on the orders of James IV of Scotland
in thanks for being saved from drowning in one of the deep
pools or steills in the Tweed, possibly following his failed siege
of Norham Castle in 1497. Its unusual all-stone construction
was said to have been to enable it to resist "fire and flood" and
so last forever. It is more likely, however, that James built it as
a secure lookout post to allow him to keep an eye on the English defenders of Norham. Whatever its original purpose, even
before it was completed, James was dead, cut down with the
majority of the Scottish nobility and ten thousand Scottish soldiers on the battlefield of Flodden, following yet another of his
opportunistic invasions of England.

But James IV was not alone in desiring a permanent memorial. From the Stone Age onward, ruling elites have attempted to build structures that preserved their eternal memory; and like James they have almost invariably chosen to do so by replacing the obvious building material, wood, with stone. As we shall see, though, this attempt has rarely been wholly successful, and architects have had to return to wood again and again to help support their buildings and make them habitable.

Even back in the Neolithic period, people were becoming dissatisfied with the ephemeral nature of wooden buildings. The large wooden halls they built rotted away at ground level after only thirty years or so. They abandoned them as dwellings, and instead, they converted many of them into houses of the dead; they used them as mausoleums in which they deposited bodies, then covered the halls with earth, to form so-called long barrows. Cat's Brain in the Wiltshire ceremonial landscape, which dates to around 3600 BC, is one such long barrow; the shadows of the postholes and of the imprint of the wooden walls in the earth are the only testaments to the barrow's original design. Later on, people constructed purpose-built tombs. They built the chambers with large upright slabs of stone, topped them with stone lintels, and covered the whole construction with earth to produce passageway tombs, chambered tombs, and smaller single-chambered tumuli. These Neolithic tombs are dotted across northwest Europe, many having lost the covering of earth to reveal the stark stone structures called dolmen that puzzled early antiquarians. The passageway tombs reached a peak of sophistication in huge Irish structures such as Newgrange, in county Meath, which dates from 3200 BC. The building consists of a huge circular mound, 250 feet in diameter and 39 feet high. A stone-lined passage extends 60 feet into the structure and ends in three chambers, each with a corbelled roof, where the dead could be deposited.

It proved fairly easy to construct stone buildings if they only had to house the dead. Unlike buildings for living people they did not need to be wholly watertight or have rooms that are large enough to walk around in. The builders could get away with low, narrow passages and small chambers that could be roofed with rough stone slabs. The idea of making ceremonial mounds, whether they contained graves or not, took off all over the world. In Britain, the fashion peaked around 2400 BC with the construction of Silbury Hill, Wiltshire. This huge conical monument is not just a simple pile of earth, but a cleverly planned stone structure; it was built using a single spiral of dry stone walls, which created gabions that were then filled with chalk rubble. The whole structure was finally waterproofed with a clay surface. Silbury Hill is huge, 128 feet high and 2 acres in extent, but it is dwarfed by the largest of the pyramids that were built in many other areas of the world: Mesopotamia, Mexico, the Andes, and, most famously, Egypt. The Great Pyramid of Khufu, the world's tallest building for over thirty-eight hundred years, is 481 feet tall and still staggers the imagination of all who visit the site.

These huge structures were largely made of stone, but their engineers still needed wood to cut the blocks. Quarry workers would first use mallets to carve a shallow groove into the stone along the desired line of fracture. They then placed radial wedges of dry wood into the crack and poured water on them. As the wood expanded tangentially, it split the stone, freeing the block and carving it into shape with little expense of energy. This technique would have been particularly important in Neolithic Europe, and in the Americas, where there were no metal tools to cut through the stone. The Egyptians also used cramps, I-shaped pieces of wood or metal, to join adjacent blocks of stone. They simply carved I-shaped impressions into the stones and inserted the cramps. The technique was most frequently used to key the bottom rows of blocks to the native rock, so forming a solid foundation for the whole structure.

In Britain late Neolithic people also developed another form of earthwork that could be impressive while demanding far less work and less complex engineering skills—henges. These are simple ring-shaped structures consisting of a raised bank with a ditch inside, but they could be built on a huge scale. The most impressive must surely be the little-known Thornborough Henges, near Ripon in Yorkshire, England, a group of three rings arranged in a crooked line replicating the arrangement of the stars in Orion's Belt. Each ring is over 250 yards in diameter and 9 feet tall, and the belt is over a mile long. When first built, the banks were probably coated in the white gypsum that underlies the area, so the rings must have been a spectacular sight. Possibly as a way of making smaller henges look more impressive, people started adding vertical members along the top of the bank, or more commonly, within the henge. The remains of postholes in some, particularly the henges that lie dotted around the great ceremonial landscape of Wiltshire, show that these uprights were at first made of wood, like the totem poles of the Pacific Northwest Americans. The best known of these structures is the unimaginatively named Woodhenge, which is sited just two miles northeast of its more famous neighbor, Stonehenge. With an overall diameter of about 360 feet, the henge is not huge, but inside it contains six concentric rings of postholes, the largest being about 140 feet in diameter. Of course none of the posts have survived, and since the Ministry of Works simply indicated their position in the 1950s using short concrete stumps, Woodhenge must be one of the most underwhelming tourist attractions in the world. In its heyday, though, Woodhenge would have been an impressive structure. The third ring of postholes from the outside were wider and deeper than the rest, suggesting that they held taller and thicker posts that supported a central ridgepole. Woodhenge has consequently recently been reconstructed not as a confusing array of vertical posts, but as a large ring-shaped building surrounding an open central area that contained ceremonial stones, like the bluestones of Stonehenge.

The position of Woodhenge, to the east of Stonehenge, and part of a ceremonial landscape that includes another large wooden henge, Durrington Walls, just 250 feet to the north, has recently led archaeologists to suggest that the wooden henges acted ceremonially as the realm of the living. This would contrast with Stonehenge, and other stone structures, which would have represented the realm of the dead. However, Stonehenge too might once have been largely a wooden structure. The central bluestones and the rings of sarsen stones that surround them are surrounded in turn by several rings of postholes, just as at Woodhenge. The stones could therefore originally have been covered or surrounded by a huge ring-shaped wooden building. The archaeological blogger Geoff Carter has even suggested that the sarsen stones themselves and their lintels could have acted as load-bearing structures in a huge wooden temple that completely covered the bluestones. The idea seems to me to be persuasive; otherwise ceremonies would have been open to the elements, a real problem on this bleak, windswept plain.

It is, of course, possible to make buildings entirely from stone that both stand up and provide shelter from the elements to living people, not just corpses. Stone buildings that are circular are particularly successful. The tiny circular trulli of Puglia, Italy, and the beehivelike cells of the Skellig Michael Monastery, off the west coast of Ireland, both make use of overlapping blocks of stone to form corbelled roofs that are rainproof, while at the other end of the scale Hagia Sophia in Constantinople and the Pantheon in Rome were roofed by huge stone and concrete domes. In a circular building, the roof stones interlock and support one another. However, it is much harder to use stone to roof rectangular structures. The Mycenaean Greeks did use corbelling to build rectangular buildings at Mycenae and Argos, but to make their buildings stand up they used massively thick walls and huge stone lintels. So when the ruling elites of Europe, the Middle East, and southern Asia sought to replace wood with stone to build more permanent and impressive temples and palaces, they all faced the same problem: how to build structures that were stable but still affordable. As we shall see, their architects usually solved the problem by hiding wooden structures within the stone ones. The history of architecture can be seen as the development of increasingly effective techniques to harness timber to stabilize and shelter ostensibly stone buildings.

Of all places, it was in the deserts of Egypt where the weather presented the fewest problems to architects. The ordinary people built their houses with simple mud-brick walls, and with flat roofs that were held up using beams made of the midribs of palm leaves. The first temples would have been grander, held up with pillars made of the trunks of palm trees or bunches of papyrus stems. To make equivalents in stone, the architects simply replaced the mud-brick walls with stone ones and carved stone replicas of the tree trunks and papyrus bundles. They soon realized that this technique freed them to scale up the building and make the pillars larger than any palm tree. Fortunately for architects, replacing brick walls and wooden pillars with stone ones presented few structural problems; the weight simply loads the stone in compression along its axis, and vertical walls are stable, especially if they are built to taper toward the top, as in the great gateways or pylons of Egyptian temples. The only problem was roofing. Because of their low toughness and tensile strength, slender stone lintels or beams stretched between two pillars will tend to crack along their lower edge and break under their own weight. Egyptian temples therefore had to be built with thick stone roof plates. Even so, their pillars had to be placed close together to reduce the stresses on the lintels, giving a cluttered space beneath. And some parts of their temples were simply left open to the elements, since in this driest of climates few ceremonies would ever be rained out.

Classical Greek temples, such as the Parthenon, were also essentially copies in stone of earlier wooden temples. Like the Egyptians, the Greeks replaced mud-brick walls and tree-trunk pillars with stone, but they had the additional problem of making a stone roof that could provide a waterproof cover over the whole structure, and this became increasingly difficult as their temples increased in size. Essentially their architects cheated and hid wooden beams within the roof structure. In Greek temples, the structure is bounded by a rectangular colonnade of pillars that holds up the edges of the shallow-pitched roof, which rises to a ridge running from the front to the back of the temple. The roof, which was covered in tiles, looks from the outside just like that of a modern house, but the engineering was

extremely primitive. Rather than being supported by an efficient truss, the wooden ridgepole was held up from beneath by
short props that rested on a lintel, a beam that was supported
at either end by internal walls or lines of pillars within the temple. This arrangement loaded the lintels in bending, meaning
they had to be heavy and clumsy, and limited the maximum
distance they could safely span between pillars. Consequently,
though Greek temples look impressive from the outside, inside
they were dark and claustrophobic, with lines of pillars bound-
ing a narrow central space. Even in the Parthenon, the widest
gap between supports was only around thirty-six feet. Indeed,
thirty-three to thirty-six feet seems to have been the maximum
span that simple wooden lintels could safely bridge in early
stone buildings. The book of Kings, for instance, tells us that
Solomon's Temple was sixty cubits long by twenty cubits wide,
around ninety feet by thirty feet, its roof being held up by cedar
beams purchased from Hiram, king of Tyre, presumably the
largest structural elements that were available.

It was the Romans who first succeeded in designing buildings
whose roofs spanned truly impressive distances. As we saw in
chapter 7 they did this by inventing wooden roof trusses in
which the outward thrust of the roof rafters is held by horizontal tie beams. In their basilicas and early churches, this enabled
them to cover naves well over sixty-five feet wide, while they
opened up access to the aisles by punching holes in the side
walls of the nave using their other major structural innovation,
the arch. The side aisles were then covered with simple lean-
to roofs. Roman basilicas were huge; the Old St. Peter's Basilica, which was built by the Emperor Constantine in around

AD 320 and which lasted into the sixteenth century, had a nave around eighty feet wide. The most impressive surviving Roman trussed roof is that of the portico entrance of the Pantheon, which supports a span between the central pairs of pillars of around forty-six feet. The original truss is said to have been covered with bronze, but this was removed in 1625 by Pope Urban VIII and cast into eighty cannons, a piece of philistinism that led contemporaries to comment wryly *quod non fecerunt barbari fecerunt Barberini* (what the barbarians did not do the Barberini family did). As the Romans sometimes also covered the aisles of their buildings, and sometimes even the nave, with barrel-vaulted ceilings, we nowadays tend to emphasize the importance of the arch. However, it's worth remembering that above their vaulted ceilings there was a wooden roof truss that actually kept out the weather.

The Roman structural designs were adapted and developed by medieval masons, and it is generally agreed that the Gothic churches and cathedrals that they built with their new, pointed arches were masterpieces of engineering in stone. Once again, however, without wood they would never have stood up or been watertight. This is most obvious in smaller parish churches, which almost always have simple wooden roofs, built by carpenters and supported by trusses of varying design; they were essentially ecclesiastical versions of the roofs of tithe barns. Light and economical to build, they proved immensely successful.

Only in the grand cathedrals did architects go all out to give the impression that their buildings were made entirely of stone. They covered the nave and choir with the soaring stone vaulting that is the celebrated glory of the Gothic style. Vast amounts have been written extolling the engineering genius behind them; the key to their success was that the outward thrust that the heavy vaulting exerted on the walls of the nave was transmitted into flying buttresses outward and eventually downward through the short walls of the aisles to the ground. But impressive as they are, Gothic vaults are nothing more than glorified ceilings. Any guided tour up to the roof of a cathedral will show that the actual roof above the stone vaults is supported by giant wooden trusses made of huge tree-trunk-size beams clumsily jointed together, which hold up the tiled or leaded roof cladding. Unfortunately, the space between the vaults and the roof acts as a vast corridor along which fires can travel: one reason why the recent fire at Notre-Dame Cathedral in Paris spread so rapidly along the nave and came so close to destroying its famous bell towers. The Italians were never so enamored of vaulting as the Northern Europeans and considered flying buttresses to be ugly, dishonest structures. In many of their great churches and cathedrals, and especially those built by the Franciscans, they replaced the stone vaulting with ceilings supported by elegant wooden roof trusses, which are plainly visible. In the great Basilica of Santa Croce, Florence, the nave is sixty-six feet wide and rises a hundred feet to the gaily painted ceiling; without the need for the heavy external buttressing, the interior is light and airy.

In England, the masons and carpenters developed a different way to reduce the need for elaborate flying buttresses. The flat arches and elegant fan vaulting of the Perpendicular style, which they developed in the late fourteenth century, enabled them to attach their wooden roof trusses to the walls just a few feet above the stone vaults, reducing the bending forces exerted on the stone. In the masterpiece of the Perpendicular, King's College Chapel, Cambridge, the single nave is ninety-five feet tall and forty feet wide, and its huge glass windows are largely unobstructed by the slender external supports. Throughout Europe, many architects did without the stone ceilings altogether and just used wooden structures. In York Minster, one of the largest Gothic cathedrals in Northern Europe, the stone vaulting above the forty-nine-foot-wide nave is a fake. It is actually wood painted to look like stone, but having only a tenth of the weight, while the roof above it is supported by an efficient wooden scissor truss. This imposture was laid bare on July 9, 1984, when a lightning strike set the roof of the south transept ablaze, causing $2.85 million worth of damage. At the time some saw this as a sign from God, since it occurred just three days after the consecration in the minster of the controversial bishop of Durham, who denied the physical resurrection of Christ. If so, God's aim was some miles too far south, as by the time of the fire the bishop had returned to Durham. Other lightweight wooden roofs were built more openly. Ely Cathedral, at the edge of the Cambridgeshire fens, is famous for its wooden octagon, which spans the large central crossing; this beautiful structure replaced a failed stone tower that was simply too heavy for the soft soil foundations.

There is one final part of Gothic churches where the low tensile strength of stone proved to be a major problem: the spire. In such a tall, slender structure, wind forces can put the windward side of the spire into tension, causing stones to be prized apart and the structure to fail. This difficulty was solved in various ways. In Germany, many cathedral spires are open filigreed structures that let the wind through, so reducing the force on them. A simple alternative is to make the spire from wood. This works well; but as spires are exposed to the elements, they are liable to warp. After centuries standing up to the Derbyshire drizzle, the spire of the Chesterfield church has been deformed into its celebrated twisted corkscrew shape. And as we have recently seen at Notre-Dame Cathedral, Paris, wooden spires are far more vulnerable to fires. But perhaps the most ingenious solution to the problem of constructing a stable spire was that adopted by the builders of Salisbury Cathedral, Wiltshire, the tallest structure in England well into the twentieth century. They used a heavy internal wooden scaffold to help them construct the stone spire, but as they completed the structure, instead of dismantling the scaffold, they hung it via iron connections from the topmost stones. The weight of the scaffold preloads the spire in compression, thereby stabilizing it; recent events show that this seems to have made this four-hundred-foot-high landmark irresistible to Russian military spies.

Architects of secular buildings went for rather different approaches to support their roofs. Many of the great halls of England were roofed with huge oak trusses. At first, most of them were built as aisled buildings like tithe barns, with a central truss supported by two rows of pillars on either side of a central "nave" and lean-to roofs above the aisles. Later on, however, carpenters left the two lines of pillars out to produce spectacular "hammer-beam" roofs; these structures, a crisscross of collars and ties, seem to float above the projecting hammer beams, apparently supported by the angels carved on the decorative bosses at their ends. This impression is emphasized by the curved "arch" of wood that rises from the hammer beams to the center of the roof. In truth, though, these hammer-beam roofs are simply rather heavy and overengineered trusses, which, because of their flexibility, exert quite large outward thrusts on the walls. The most celebrated example is Westminster Hall, the oldest remaining building of the Palace of Westminster. The hammer-beam roof covers a breathtaking open area sixty-eight feet wide, but the outer walls of the hall had to be propped up from the outside by huge buttresses to stop them from being pushed over.

But perhaps the most spectacular of the wooden-roofed buildings of the Middle Ages is the Palazzo della Ragione in Padua, Italy. The roof of this medieval market building is beautifully curved, resembling the hull of an upturned ship, and covers an area 260 feet long by 90 feet wide. And for the wood enthusiast there is an added bonus of a huge twenty-foot-high wood sculpture of a horse, though quite why it is there or how people got it in is a mystery not explained by any of the tourist guides.

So far we have concentrated on public buildings. However, as the wealthy and powerful started to build larger private dwellings out of stone, other problems emerged that could only be solved by using wood. Since castles and the palaces and mansions that succeeded them are usually several stories high, the builders had to insert wooden beams into the walls to support the upper floors. You can spot the beam holes when you visit ruined castles; they are always punched into the walls just below the level of the fireplaces that seem to hover on the walls. The builders also gave their palaces lightweight wooden roofs. The great Venetian architect Andrea Palladio was the first to produce designs for simple, efficient roof trusses, and his methods were copied all over Europe. But as Palladian mansions were built farther and farther north, a new difficulty became apparent: the cold. Stone buildings are ideal for the climate of Italy, where the high thermal capacity of the stone keeps them cool on hot summer days and maintains an equable temperature. In the cold, damp climate of Northern Europe the high thermal conductivity of stone meant they lost heat rapidly in winter, and the high thermal capacity meant that once they got cold, they took ages to heat up again. You only have to go to a Christmas carol concert in a stone church to appreciate the problem: churches are notoriously cold in winter, so you should always wrap up well. In the Middle Ages, nobles overcame to some extent the problem of living in cold stone castles by furnishing the inside of their walls with tapestries. In later palaces and mansions, architects replaced the tapestries with wooden paneling. Wood is a far better insulator of heat than stone, largely because of its cellular nature; the innumerable tiny air spaces restrict heat flow. Not only is wood ten times as effective as stone at stopping heat loss, but because the paneling is attached to the stonework via battens, this leaves an additional air space, which provides further cavity insulation. Perhaps the coziest room of all in country houses was the library, which was insulated not only by the paneling and bookshelves, but also by the books themselves. Wood was even used to insulate the windows. In Southern Europe, external louvered shutters kept out the energy from the heat of the sun, while allowing some cooling airflow. In Northern Europe, the windows were instead furnished with folding internal shutters that could be closed over the windows at night to conserve heat and opened in the day to let in as much sunlight as possible.

There is one final way in which wood supports stone buildings—from below. In chapter 7 we saw that wooden buildings are vulnerable to rot if their structure is allowed to touch the soil surface, where the wood can be alternately wetted and dried out. However, wood is not only stable when it is kept constantly dry; it also resists rot well when it is kept permanently wet, because the fungal hyphae that break down the wood cannot survive in anaerobic conditions. This is also the reason why living trees stay free from fungal diseases; their cells are filled with water. Only when a branch dies and starts to dry out does it become vulnerable to fungal attack. Many of the churches that were built in the peatlands of Northern Europe, such as Ely Cathedral, which was built in the fens of East Anglia, were therefore built on foundations made from wooden rafts that were sunk into the wet peat. The same is true for many of the great cities of Europe. In Venice, Amsterdam, and Hamburg, most of the buildings are supported on wooden piles. In Venice, the elm logs used for this purpose can be sixty feet long; huge arrays of them were driven down through the soft silt to the hard clay below, before being leveled off just below the water table, covered in wood planks, and finally in brick foundations. Supported from below, sheltered from above, and insulated from within, the stone and brick buildings of Europe have always been dependent for their stability and comfort on wood.

But though stone buildings took over from wood in Europe and much of central Asia, this was certainly not true in the Far East. The temples and palaces of China and Japan continued to be built in wood, including the largest palace complex ever built, the Forbidden City in Beijing. Chinese buildings have traditionally been regarded by Western engineers as fairly primitive structures, since they do not incorporate the trusses that Western architects use to support the roofs of their buildings. Instead of being triangulated to make the structure rigid, all the beams in a Chinese building are at right angles to each other. In a typical temple, for instance, the pillars at the front and back of the building simply support a large horizontal beam. This supports a shorter beam above, and in turn this supports a still shorter beam. In this unusual-looking stepped arrangement, the heavy roof is supported via brackets at the ends of each of the beams. This design means that the beams are all loaded in bending and have to be thick and heavy to withstand the weight of the roof. However, the arrangement does have two advantages. First, it enabled the architects to build graceful curved roofs, something that would not have been possible if they had used a triangulated truss structure. Second, the main benefit, though, was structural. The pillars and beams are all joined to the members above them and to the roof via complex wooden brackets called dougongs. These are made up of many interlocking wooden joints, each of which is interlocked at right angles to the next; together this makes up a loose mechanism that takes any amount of weight and acts just like the shock absorber of a car. The reason for this complexity becomes apparent when one thinks about the tectonic conditions in China and Japan: both countries border the Pacific Rim and are prone to massive earthquakes. Recent research by Chinese engineers shows that the flexible design of the buildings and the shock-absorbing dougong joints protect them from damage. As an earthquake hits, and the ground moves around, the pillars sway, while the inertia of the heavy roof keeps it still and the energy is absorbed within the dougong joints. Tests on models have shown that these buildings can withstand shocks that reach over 10 on the Richter scale, more powerful than any earthquake yet recorded. It's no surprise then that the Forbidden City has stood for six hundred years. Many temples in Japan are even older. The pagoda of the Horyu-ji Temple dates from around AD 600 and like other pagodas has a central wooden pillar, whose flexibility provides further protection from earthquakes. As it swings, it absorbs the energy of the quake and prevents damage elsewhere in the structure. With the advantages of such advanced seismic design, East Asian priests have just come to accept the need for greater vigilance against fires, and the need to replace wooden beams every few centuries!

Paradoxically, therefore, apparently ephemeral and primitive wooden buildings can survive where stone ones are destroyed. Standing in the famous Ramesseum, the funerary temple to Ramses II on the west bank of the Nile, and seeing how centuries of subsidence have turned the stone temple and its statues into a colossal wreck, one is tempted to agree with Shelley. Commenting on man's folly in trying to make permanent memorials in stone, we might well say, "Look on my works, ye Mighty, and despair."

CHAPTER 10

Limiting Our Outlook

James IV of Scotland not only had pretensions to construct eternal buildings, but also grandiose plans to recapture Palestine for Christendom; and to do so he ordered the construction of a fleet of thirty-eight warships led by an enormous flagship. The *Michael* was to be the largest ship in the world, a four-masted carrack, 240 feet long, 36 feet wide, and with a displacement of around a thousand tons. Construction posed logistical problems, though, since Scotland had no shipyards big enough to build it in, and James had to found a new port for this purpose, Newhaven, two miles north of Edinburgh. The ship took five years to complete and was said to have consumed "all the woods of Fife," but she was eventually launched in 1511 and fitted out by 1512 with twenty-four heavy guns with which to fire broadsides. The achievement put Henry VIII of England, that most egotistical king, into a fit of envy so strong that it was not assuaged until he had built an even bigger ship, modestly named the *Henry Grace à Dieu*, which had two gun decks sporting forty-three heavy guns and a total displacement of around fifteen hundred tons. Neither ship had a distinguished career. James was diverted from his plans for a crusade, by his commitments, under the Auld Alliance with France, to go to war with England to divert Henry from his war with Louis XII. As we have seen, he was killed at Flodden Field after his disastrous invasion of 1513, and his big ship was sold off to France at a bargain price of forty thousand livres. Henry's ship, meanwhile, proved top-heavy and unstable and had to be remodeled into a smaller ship, which was mostly used as a diplomatic vessel. Its most important task was to take Henry to the famous Field of the Cloth of Gold in 1520, where he memorably (to the French if not the English) lost an impromptu wrestling match against Francis I.

What these events show in hindsight and with the benefit of a broader historical perspective is not only the hubris and folly of powerful men, but the stasis of technology in the High Wood Age. For these two ships, large though they were, were smaller than the great ships of antiquity. For instance, in 240 BC the tyrant of Syracuse, Hiero II, built an enormous three-masted barge, the *Syracusia*, the world's first cruise ship. Only the lowest of its three decks was built to hold cargo, around seventeen hundred tons of the grain that was one of Sicily's biggest exports. The middle deck was fitted out for passengers and included thirty cabins, a chapel, library, gymnasium, and bath complex, all elaborately decorated and furnished with marble surfaces, paintings, statues, and living foliage. Finally the top deck housed a unit of marines and their equipment, making the ship both luxurious and secure. Like the great British engineer Isambard Kingdom Brunel's *Great Eastern* in the nineteenth century, one main difficulty proved to be launching this huge vessel—fortunately Hiero could call on his resident engineering genius, Archimedes, to help with this—while the huge mainmast was transported from the mountains of the toe of Italy by the engineer Phileas of Tauromenium. Unfortunately the ship proved too big for most ports and it undertook only one journey, to Alexandria, where it was presented as a gift to Ptolemy III and renamed *Alexandreia*.

Hiero's ship was plainly a vanity project, like the *Michael* and *Henry Grace à Dieu*, but the Romans later built entirely practical ships of similar and even greater size. The majority were used to transport the vast amounts of grain produced in the breadbasket of Egypt twelve hundred miles across the Mediterranean to the imperial capital, Rome, and could carry cargoes in the range of a thousand to twelve hundred tons. We are fortunate that one of these ships was blown off course to Athens in the second century AD, where the writer Lucian wrote up an enthusiastic account of it:

> What a size the ship was! One hundred and eighty feet in length, the ship's carpenter told me, the beam more than a quarter of that, and forty-four feet from the deck to the bottom, the deepest point in the bilge. What a mast it had, what a yard it carried, what a forestay held it up! The way the sternpost rose up in a gradual curve, with a gilded goose head set on the tip of it, matched at the opposite end by the forward, more flattened, rise of the prow with the figure of Isis, the goddess the ship was named after, on each side! And the rest of the decoration, the paintings, the red pennant on the main yard, the anchors and capstans and winches on the foredeck, the accommodations toward the stern—it all seemed like marvels to me! The crew must have been as big as an army.

The Roman emperors built even larger ships to transport obelisks from Egypt to decorate the environs of Rome. The obelisk that now stands in front of St. Peter's is around eighty-two feet high and weighs around five hundred tons. To bring it over to Rome in AD 40, the Emperor Caligula constructed a huge vessel that was ballasted with eight hundred tons of lentils, making a total load of thirteen hundred tons.

These Roman ships were only matched in size in the seventeenth century by the largest Manila galleons of the Spanish empire and the British East Indiamen of the eighteenth century. Roman shipwrights appeared to have reached the limits of construction in wood. And this was not the only field in which the Romans excelled. Roman architects seemed to have reached the limits of the span of wooden roof trusses, around eighty feet. Carts and wagons got no bigger after the classical period, and the design of wheels remained unchanged from the start of the Iron Age well into the nineteenth century. Carriages and chariots got no faster over the same period. Wooden houses remained unchanged in design, as did furniture, and farm tools such as plows. And after the Roman period, few improvements were made to woodworking tools until the development of sawmills in the sixteenth century. Nor did people come up with any major changes in the way they used wood to manufacture other materials—metals, pottery, glass, or leather—or in its industrial uses in saltworks, soapmaking, dyeing, or the production of alum. At the end of the early modern period, towns were no bigger than they had been in antiquity, the population density, and the output of agriculture, no higher. It's no surprise, therefore, that Europeans looked back to the ancients as superior beings and viewed the learning of the Ancient World in awe; they lacked confidence in their own ability to improve their condition.

As we shall see, the reasons for this were not as obvious as you might suppose; our reliance on wood influenced not only the engineering of the period, but also the very structure of preindustrialized societies. Just as wood had shaped our bodies and minds, so it shaped our civilization. In doing so it imposed limits that hindered what we now think of as progress. It stifled economic and population growth and the development of technology and science, and it helped impose a worldview that was essentially conservative.

Wood was such an essential component of everyday life that one might expect a limit to wood production was responsible for the lack of progress. As we have already seen, almost all the possessions of everyday folk were wooden, while those that were not actually made of wood needed large quantities of wood to produce. In the Middle Ages, around thirty pounds of wood were needed to smelt one pound of iron, for instance. People burned wood in even greater quantities to cook food and heat their houses, and wood was a vital energy source for the major industrial processes of the age: salt making, brewing, tanning, fulling, and dyeing. In the language of chess, wood was an overworked piece. And as the human population grew, and land was cleared for agriculture, forests would have been grubbed up, reducing the wood supply still further. You might well think that this would eventually have led to a shortage of wood, stunting further material progress. Popular histories, after all, are full of stories of how using wood led to deforestation and disaster.

However, it is necessary to check whether such stories stand up to scrutiny, and one of the best ways to do that is to carry out rough "order of magnitude" calculations; in this case we need to compare wood use and potential wood supply. Recently, economic and environmental historians have started to use this approach with enlightening consequences. For instance, Paul Warde of the University of East Anglia has calculated that in England and Wales in the 1650s, people obtained around twenty terajoules of heat energy per year by burning firewood, slightly more than the energy expended by the metabolism of the people themselves, and that expended by their farm animals. Since burning wood produces about 7.3 megajoules per pound, this translates to around 1.2 million tons of firewood being burned each year. This seems like an enormous amount, but foresters have found that coppice woodland can produce about 2 tons of wood per acre per year, so it would require only six hundred thousand acres or 950 square miles of coppiced woodland to produce this amount of wood— just 1.6 percent of the total surface area of England and Wales. But what about the land area needed to supply timber?

Nowadays, only 40 percent of the wood harvested globally is used for nonfuel uses, and if the proportions were similar in preindustrial times, this would equate to around 0.88 million tons of timber. Timber trees lay down wood rather more slowly than coppices for two reasons: first, because it takes many years for their canopy to develop, and second, because as they grow taller and reach maturity, the difficulty in transporting water to their canopy means they have to shut their stomata and stop photosynthesizing earlier in the day. Consequently the growth rate of high forest is roughly half that of coppice, around a ton per acre per year. Even so, the demand for timber could be met by an area of woodland of around fourteen hundred square miles. This makes a grand total of woodland needed of just 4 percent of the land area available. Yet we know that the total woodland cover in England and Wales was actually something in the region of 10 percent in the preindustrial period. The answer is therefore clear; we cannot put the lack of economic growth down to the failure of trees to produce wood fast enough, even in England, which was probably the least forested and most densely populated country in Europe.

However, though trees could have grown fast enough to supply all the fuel and timber needs of preindustrial countries, even densely populated ones with low forest cover such as England, there would have been a problem in actually cutting and transporting the wood to where it was needed. In forests, wood is evenly spread over huge areas—mature coppice only contains around forty tons of wood per acre—and it is not an energy-dense fuel. Dry wood contains only half the energy per unit mass as coal and is only around 40 percent as dense, so it contains only a fifth of the energy per unit volume. It is extremely time-consuming to harvest, cut into usable pieces, and pack efficiently into a small space. English woodsmen traditionally cut coppiced firewood into small relatively straight twigs and bound them together into faggots three feet long and with a diameter of around eight inches. Only then could the wood be transported on wagons away from the forest. The difficulties in then getting wood to its destination were great, particularly away from navigable waterways. The poor state of roads in preindustrial times made wheeled transport slow and expensive, increasing the price of the wood to exorbitant levels if it had to be moved more than a few miles.

This would not have been a major problem for villages and small towns, where wood could easily be sourced locally, but it would have become one for larger towns and cities. Ad van der Woude from the Agricultural College of Wageningen, Netherlands, and his colleagues estimated that the town of Odense in Denmark, with a population of five thousand people required around seventy-five hundred tons of wood per year. This could have been provided by exploiting six square miles of woodland, and if the woodland cover of the area was around 20 percent, allowing market gardening to use the rest of the land to supply the town's food, this meant exploiting a surrounding area of thirty square miles, giving an exploitation zone that was only around six miles in diameter: easily accessible to the town, even by road transport. However, a town of fifty thousand people, ten times as large, would have needed a zone around twenty miles in diameter, and one of five hundred thousand a zone sixty miles in diameter, far too great for easy access along the poor roads of the Middle Ages.

It's not surprising, therefore, that in medieval Europe it was only ports, and cities on large navigable rivers, that grew to any reasonable size. The largest city on the Continent, Paris, which had grown to around four hundred thousand inhabitants by 1600, obtained the majority of its firewood supplies in the form of coppiced beech that was harvested from the Morvan mountains of Burgundy. The wood had to be floated down the rivers Yonne and Seine, a distance of over 120 miles to the city. So important was the trade that the whole Seine River basin had to be adapted to allow rafting of wood. Other large continental cities were also supplied with wood in the same way, and the most impressive supply line for firewood was the Rhine. It acted as the conduit to transport softwood, cut from the hills of the Black Forest in southwest Germany, down to the cities of Strasbourg and Cologne and to the urban areas of the Netherlands. Foresters built wooden chutes to allow tree trunks to slide downhill into the Rhine's tributaries. The logs were then built into rafts that could be floated down the Rhine. The rafts could be extremely long, each section of logs being joined to the ones in front and behind with specially twisted wooden ropes, allowing the rafts to bend around the curves of the river. They were looked after by whole families, who steered them with poles and

At 450,000 years old, the Clacton Spear, discovered in Essex, England in 1911, is the world's earliest known wooden artifact. The point was sharpened with a stone blade, either when the wood was still green or after being charred in a fire. Archaeologists have interpreted the artifact's intended use in myriad ways: it could be the broken end of a digging stick, lance, or a spear.

1

The ability to fell trees enabled Mesolithic people to build roomy round houses. The turf covering this reconstruction of an eight-thousand-year-old hut in Howick, Northumberland, hides a complex structure underneath, with a ring of posts supporting the rafters.

2

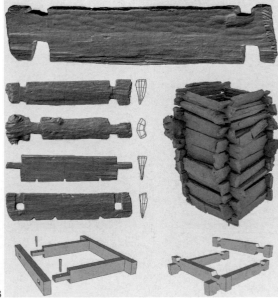

The first carpentry? In this 3-D rendering of a 7,300-year-old well lining from eastern Germany, the bottom of the frame features mortise-and-tenon joints, while the upper layers are joined by interlocking grooves. The rough ends to the planks show the difficulty Neolithic people had cutting wood across the grain before the advent of metal tools.

3

The tool innovations of the Neolithic LBK people allowed them to build long, narrow timber dwellings for multiple families. This reconstruction of a LBK longhouse from La Hougue Bie Museum in Jersey has a roof supported by five lines of poles, the outermost ones forming the walls, which are all dug into the ground, while the hurdle door is made from coppiced poles.

4

The 5,200-year-old Ljubljana Marshes Wheel is the world's earliest surviving wheel and axle. The wheel is made from two planks joined by a series of battens (now broken) that fit snugly into grooves chiseled out of the planks. The square hole in the wheel fitted onto the axle and rotated with it.

5

The reconstructed 4,500-year-old solar ship of Khufu, now in a museum next to the pyramids and sphinx at Giza, Egypt. Sealed disassembled in a pit at the foot of the Great Pyramid, the 143-foot long vessel was built to take the resurrected pharaoh across the sky. The short cedar planks were joined together with mortise-and-tenon joints as in many Bronze Age boats.

6

7 Statue of Ka-Aper, around 4,500 years old, from the Cairo Museum in Egypt. Carved from sycamore, the statue presents the priest with a distinctive personality, in contrast to the idealized stone statues of the pharaohs. One can even see his receding hairline.

8 The prow of the early-ninth-century Oseberg ship from the Viking Museum, Oslo. Note the fine carving, graceful lines, and clinker construction of the hull. The individual strakes were riveted together with iron nails before an internal frame was added to strengthen the hull.

9 Detail of the Bayeux Tapestry from the eleventh century, showing the construction of William the Conqueror's fleet for the Norman invasion of England. On the left, foresters fell trees, while in the center a carpenter shapes planks using a broadaxe. On the bottom right, workers shape the hull of a ship, while at the top right the shipwright assesses the alignment of the hull and a workman drills holes in it with an auger.

Heddal Stave Church, built in the early thirteenth century near Telemark, Norway. In stave churches, the walls are made from lines of vertical split logs, or staves, while the main structure is supported by a framework of logs and beams and the roof is covered with wooden shingles.

10

Harmondsworth Great Barn was built near London in the early fifteenth century and could store three thousand tons of grain. The oak pillars are mounted on stone supports to prevent the base rotting, while the roof is held up by a series of rough oak trusses.

11

Esslingen Town Hall, in southwest Germany, built in the early fifteenth century, is a typical wood-framed medieval building. The upper stories jut out from the ones below, helping prevent the floors from sagging, while the diagonal cross struts stabilize the structure against the wind.

12

The hammer-beam roof of Westminster Hall, London, was built in the late fourteenth century and is the oldest part of the Houses of Parliament. The impressive-looking structure spans sixty-eight feet but the truss is actually rather inefficient. The curved elements do not act as arches but are merely decorative, and the walls have had to be buttressed from the outside to prevent collapse.

13

A brightly painted dougong, or roof bracket, from the memorial archways at the Mogao Grottoes, Dunhuang, Gansu Province, China. Dougongs support the roof of Chinese buildings, while their flexibility allows the building to sway harmlessly in earthquakes.

14

HMS *Victory*, here seen at the Portsmouth Historic Dockyard, England, was the ultimate wooden warship and Nelson's flagship at the 1805 Battle of Trafalgar. The huge masts are made from the trunks of American white pine trees and the hull from oak beams and planks. The complex rigging was controlled with the help of large numbers of wooden blocks or pulleys.

15

A Windsor chair made in England in the 1760s. The arms and back have been steam bent into graceful curves, allowing the chair to sport the Gothic arch that was fashionable at the time.

16

A serpent made in France in 1820. A predecessor of the tuba, the serpent was a bass cornett, and the last of the wooden "brass" instruments with a trumpetlike mouthpiece to be commonly used. Serpents formed part of church bands well into the twentieth century and were mentioned in Thomas Hardy's novel *Under the Greenwood Tree*.

17

18

Ten men constructing a balloon-framed house in the western United States, 1877. Note how narrow the timbers are compared with those in the medieval town hall in image 12. They were joined together using machine-cut nails and covered with boards inside and out.

19

A train on the trestle TVRR Bridge at the head of Fox Gulch, Alaska, 1916. To bridge canyons, Americans preferred to build trestle bridges over stone viaducts or earthen embankments. They were made by joining logs together with wrought-iron bolts to build a complex braced structure. The funnel-shaped smokestack of the train shows it was burning wood, not coal.

20

A log raft on the Columbia River, Oregon, 1902, shows the huge scale of forestry operations even before the invention of chain saws and heavy logging equipment. Log rafts have been floated down rivers to cities around the world like this for centuries.

An Albatros DVa fighter flying at Duxford Air Museum, Cambridgeshire, England. This German World War I fighter had a plywood fuselage that was formed into a beautiful streamlined shape.

21

22

The Winter Gardens in Sheffield, England, was built in 2003 and is an example of the architectural possibilities of modern glulam construction. Here each laminated larch beam has been curved into a graceful parabola, bringing new life to the conventional glasshouse. Note how the beams are supported at the base with steel brackets.

At 280 feet high, the eighteen-story Mjøstårnet mixed-use tower in Brumunddal, Norway, is currently the world's tallest wooden building. It is supported internally by a framework of glulam and CFL beams rather like medieval halls and balloon-framed houses.

23

rudders and anchored them overnight using logs that they drove into the riverbed.

Since it was difficult and expensive enough to supply cities with enough wood to keep people warm and cook their food, it would have been totally uneconomic to locate fuel-intensive industry in urban areas as well. Industry was better sited well away from cities, and in areas where trees grew well but crops did not, so that there was little competition for land. In premodern times, therefore, industrial units were largely small-scale rural enterprises, in contrast to the large urban mills of the Industrial Age. In Northern Europe, for instance, glassmakers produced "forest glass." They used potash made from burning beech trees, dissolving the ash and crystallizing out the pure potassium carbonate as the flux to add to the main ingredient, sand, and they used charcoal as the fuel for their kilns. Soap was also a forest product made by combining potash with animal fats. Gunpowder was made far from major settlements, using alderwood to produce the reactive charcoal component; this diffuse, porous wood is perfused throughout with large vessels that allow the trees to grow rapidly in waterlogged soil, and which gives alder charcoal a huge surface area that speeds up combustion. Around the salt mines of Austria, pure salt was crystallized out by heating the brine with wood cut from the forested slopes of the surrounding Alps. And in China, the main porcelain works were located in Jingdezhen, central China, nine hundred miles from Beijing, where there were not only plentiful supplies of china clay, but vast areas of woodland to power the kilns.

But it was the iron industry that demanded and exploited the greatest areas of woodland; ironworks had to be located not only where there were iron-rich ores, but where there was abundant woodland and little competition from other land uses. In England, the great iron-making area, from Roman times onward, was the Weald of Kent, Sussex and Surrey, a huge dome of exposed Cretaceous rocks, iron-rich sandstones and impermeable clays where the world's first dinosaur fossils were found in the early nineteenth century. Iron ore was smelted using charcoal made from the oak, beech, and hazel that thrived on the clay soils and was pounded into shape using hammers powered by water mills on the small streams. The whole area was an industrial powerhouse, exploited to its fullest, and using intensive coppicing to provide a regular supply of charcoal. There were few competing demands for the land from agriculture, as the clay soils were too heavy to plow and the sandstone soils too infertile to grow cereals. This explains the name of the region, as *weald* is derived from the Germanic word for a forest, *Wald*; even today the area is among the most densely wooded in the country, despite the recent influx of London commuters. Other areas that are famous for producing iron and steel also still have high levels of tree cover, the hills around Sheffield, South Yorkshire, for instance, and South Lakeland, around Lake Windermere, now far more famous for its Romantic poets. Elsewhere in Northern Europe, other centers of the iron industry included the Ardennes forest of Belgium, and northern Sweden, where the huge iron ore deposits were smelted using charcoal from the vast surrounding conifer forests and shipped out in summer from its Baltic coast.

If industrial processes were small and scattered in preindustrial Europe, the same was true of the craft industries based on wood and iron products. Woods were filled with colliers, hurdlers, turners, and bodgers. Every village had its carpenter and blacksmith, and there were builders, cartwrights, wheelwrights, and cabinetmakers in every market town. These sorts of businesses were hardly likely to grow large or to need high levels of investment; it was not a world that was conducive to venture capitalism or to industrialization.

Supply problems were just one of many factors that conspired to prevent an increase in the output and productivity of wood-based trades. As we have seen, it is a complex and time-consuming process to make even the simplest of items out of wood using the hand tools of the carpenter. The pieces take time to cut to size, and the joints take even more time to work out, measure, mark out, and cut. Further time is then needed to glue them together with the animal-based glues then available. George Eliot's great pastoral novel *Adam Bede* begins with a scene at the local carpenters', where Seth Bede, Adam's muscular but feebleminded brother, proclaims (mistakenly) at the end of the day that he has finished the door he was making. Making a door must have taken several days' work. And this is on top of the time the forester would have spent selecting and cutting down the trees, the time the sawyers would have spent cutting it up into planks, the time taken to lay down the wood for years of seasoning, and finally selecting the right planks for the job. The whole trade would have been the total antithesis of modern "just-in-time" production methods; even today, carpenters still strive more for quality and finish rather than for speed and productivity. Since it demands such a huge amount of skilled work, carpentry would have involved high labor costs, so it is not surprising that furniture was sparse, even in wealthy households, and the rare oak chairs, tables, and chests would have been expected to last for generations. More complex items would have taken even longer to make. Wheels took several days, carts several weeks, and ships years to construct. So though only a small fraction of the wood was used as timber for construction purposes, the output of wood products would have been limited by the size of the workforce available and the high labor costs.

The small size and scattered nature of the woodcrafts also helped stifle innovation and stall technological progress. Even if a craftsman made a useful innovation, it would be unlikely to spread because it would have been hard to pass on. Craft trades were handed down through the generations, largely from father to son, and the techniques were demonstrated by example, not by written or even verbal instructions. Craftsmen developed a feel for their trade over a long apprenticeship, not through textbook study. And with workshops spread evenly across the countryside, and little contact between the different wood-based trades, new skills would only slowly spread. Craftsmen would carry on working in the same way, following tradition, in other words because "if it's the way we've always done it, then that's the right way to do it." This homage to tradition is a conservative trait; it ensures that standards are maintained and mistakes averted, but it strangles innovation. Each of the wood trades also had its own guild, which would have guarded its secrets from outsiders, and even from the other wood trades, preventing diffusion of new ideas. Finally, the hands-on learning approach and the way people were taught by example would also have limited technological development, since it meant that the craftsmen had no language with which to express the behavior of wood or to understand the forces to which it is subjected. The designs of traditional woodworkers therefore lacked a sound structural basis, and the woodworkers carried on making the same basic engineering mistakes as their forebears.

The main weakness in the design of traditional wood structures derives from wood's anisotropy—the very different properties along and across the grain—which presents great difficulties to the carpenter. It is hard to join two elements while preventing the wood from splitting. We saw in chapter 7 that carpenters overcame this problem by joining the elements of their structures at right angles, holding them together with mortise-and-tenon and dovetail joints. These joints are reasonably good at withstanding axial forces—preventing the two parts from being pulled apart—especially when they are glued, so wooden structures rarely break, one reason we all fancy ourselves carpenters. The problem, however, is that though they are strong enough, wooden structures made up of a series of right-angled joints are not rigid enough. They are all too prone to deforming, especially in the mode that engineers call shear. If a square structure is pin jointed at the corners, it can readily be deformed into a rhombus, and a rectangle deformed into a quadrilateral. Most traditional wooden joints get stressed in shear and tend to work loose over their lifetime. It's why panel doors with a rectangular frame tend to drop under their own weight. It's why bookshelves lean. It's why old tables sway and old chairs creak. And it's responsible for the lack of rigidity in bed frames that results in embarrassing bangings of the headboard when they are put under dynamic stress. It's why the wheels of carts start to wobble, and the frames of carriages twist. Traditional carpentry produces inefficient structures.

At least one group of woodworkers—builders—did seem to appreciate the problem and came up with a good working solution. If you look at the internal structural timbers of a roof, you will see diagonal members incorporated between the rafters, wind braces, which prevented the rafters from collapsing sideways like a pack of cards. Similar diagonal elements can be seen in the walls of many half-timber houses as well as in board batten doors and five-bar gates. When these structures are subjected to shear, the diagonal member is loaded axially, and its resistance to being compressed or stretched gives the structure the shear rigidity it needs. However, it is unlikely that builders fully understood the forces they were dealing with, since many of the designs they commonly used for roof trusses—the queen post truss, for instance—loaded the tie beam in bending, making the structure heavier and less rigid than it need have been. And as we have seen in the last chapter, hammer-beam roofs were heavily overdesigned and inefficient structures. Most crucial of all, though, builders failed to pass their solutions on to any other trade, the most serious consequences being to shipbuilding.

In all the ship designs we looked at in chapter 7, whether the shell was built first and then stiffened by the frame, or whether the frame was built first and then coated with the shell, the structural elements are all at right angles to one another. The frames run around the hull, while the planks run along its length. This seems to be a strong design until you consider how ships are loaded when they are floating on the sea. Most of the buoyancy keeping a ship afloat occurs around the center of the ship, where it is at its widest and where the hull lies deepest below the water. In contrast the two ends of the ship are narrower, and the prow and stern actually poke out of the water. Consequently, the weight of the two ends tends to pull the bow and stern downward, and so bend the ship, a deformation known as hogging, loading the frame in shear. And in rough seas, as waves alternately raise and lower different parts of the ship, the shear forces will change. Just like the frame of a chair, the ship's structure will loosen. It will start to leak and may in some cases even spit out the caulking material that is

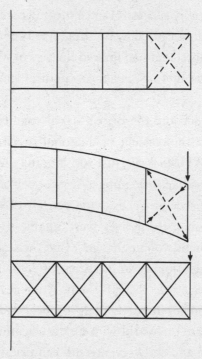

The importance of diagonal elements in a wooden structure. If there is no cross bracing (top) a square structure is easily deformed (center). Incorporating cross bracing (bottom) makes the structure much more rigid since deforming it would stretch or compress the braces.

supposed to waterproof the joints between the planks. Almost all traditional wooden ships let in quite large amounts of water and had to be continually pumped out to keep them afloat. Egyptian naval architects tried to counter this with an ingenious structural device. They used a rope running between the bow and the stern, and above the cabin, to act as a tension cable, holding up the ends of the ship and preventing hogging. This hogging truss did limit the movement to some extent and was later successfully used in Mississippi paddle steamers, but it was never entirely successful since the hull could still undergo shear deformations caused by twisting. It was not the properties of wood, therefore, that had limited the size of wooden ships to under two thousand tons, but the structural ignorance of their designers.

Not until 1805 did the British naval engineer Robert Seppings introduce the concept of diagonal bracing, including it in the design of HMS *Kent*, just too late for the Battle of Trafalgar. Sadly this had little impact, as wooden ships were soon afterward superseded by ones made entirely of wrought iron.

If the craftsmen themselves were unable to improve their mechanical designs, you might think that the students educated in the newly developing universities of Renaissance Europe would have been able to help them. After all, scholars were turning from studies in purely philosophical and religious matters to investigate the world with a new scientific outlook. But in the sixteenth and early seventeenth centuries the interests of the scientists were in pure science, chiefly in the movements of celestial bodies, in identifying the elements, and in understanding the workings of the human body. They had little interest in applied science, not surprisingly as the crafts were all carried out far away from university towns, and by an uneducated social class that jealously guarded its trade secrets. The separation of crafts from the intellectual life of Europe was yet another way in which our dependence on wood held back material progress. Not until intellectuals were forced to live cheek by jowl with craftsmen and industrialists could these groups begin to learn from one another.

PART 3

WOOD IN THE INDUSTRIAL ERA

CHAPTER 11

Replacing Firewood and Charcoal

In 1661 the essayist and diarist John Evelyn published what has often been considered a seminal environmental tract. In *Fumifugium* he railed against the smogs that cloaked London then, just as they were to do for centuries to come, correctly laying the blame on the burning of high-sulfur coal. But Evelyn's was not primarily a work of environmentalism—he did not suggest a return to burning wood—but a cross between an early proposal for town planning and the first great NIMBY rant in history. In sixty-four pages of his overblown prose, he suggested banning all industry from the City of London, and setting up a "green belt" of sweet-smelling trees, herbs, and flowers so that gentlemen such as himself would not have to put up with the smoke and stench that filled the streets. Instead, London would become more like the planned cities of the Continent, where anything disagreeable to the well-bred inhabitants was kept out of sight and mind.

Nowadays zonal planning is a common policy. It contributes to healthy and efficiently run cities, though at the same time hiding our dependence on heavy industry. However, Evelyn failed to grasp the irony that he could only have written and published his complaint because burning coal had allowed London to become the largest, fastest-growing, and most freethinking city in Europe. It was the switch from wood to coal that would soon enable it to support the most brilliant intellectual revolution Europe had seen, a revolution of which Evelyn himself was to become a player and chronicler. It was the switch that was to propel first Britain, and eventually the rest of the world, into an industrialized and urban society, and move it out of the Age of Wood, with huge environmental consequences for the planet.

We saw in the last chapter that the demand for firewood, and the difficulty of supplying cities with this hard-to-transport material, limited their size, constrained where they could be sited, and ensured that they were isolated from the industry that provided much of their wealth. But toward the end of the Renaissance, two countries in Northern Europe started to escape from the constraints imposed by the supply of firewood, utilizing two very different new sources of heat energy.

Even in the fifteenth century, the Netherlands was densely populated and had a limited local supply of wood. Much of its forest cover on its higher ground had been cleared for agriculture, while the rest of its land, which was low-lying and covered with waterlogged peat, had never been forested. This was not particularly unusual; peat was widely spread across much of Northern Europe, having been laid down since the end of the last ice age in cooler, waterlogged areas where the ground was too wet to allow the mosses, sedges, and grasses that grew there to decay. The result was a buildup of their remains—peat—which could be laid down at rates of up to an inch every twenty-five years. Peat had long been exploited all over Europe for fuel, at least on a small scale. Villagers would cut peat in early summer and allow it to dry out before carting it to their homes and burning it over the winter. However, it had never become a major fuel, largely because it was even more uneconomic to move than wood. Peat contains only half the energy per unit mass as wood and has one-fifth of its density, so it contains only 10 percent of the energy per unit volume. It could only be exploited on a large scale if it could be transported by water— which would usually have involved the expensive construction of a system of canals.

The Dutch were fortunate, however, in that two areas of low peatland, following sea level rises after the last ice age, lay at sea level. Of these the most important was that in the Holland region, an area of around 250 square miles bounded by a ring containing most of the major towns of the country: Naarden, Utrecht, Gouda, Rotterdam, Delft, Leiden, Haarlem, Alkmaar, and Amsterdam. As their trade increased in the sixteenth century, the Dutch found that they could readily exploit these peat reserves. When they dug up the peat, it exposed the underlying clay, whose surface was two or three meters below sea level, and this automatically opened up new lakes and canals along which the peat could then be transported to nearby towns. This meant that the Dutch were sitting on a whole new source of readily exploitable energy. Jan de Zeeuw of the Agricultural University of Wageningen has calculated how much energy this could supply by estimating the depth of peat and the rate at which it was cut. He found that the new low-peat reserves gave the Dutch access to twenty-five petajoules of heat energy each year, three times the energy per person than the English were then obtaining from their firewood. Moreover, in removing the peat, they exposed rich clay soils that they could then drain and convert into productive arable land.

The wealth of the Netherlands was secured; the extra energy drove the Dutch economic expansion of its Golden Age, which lasted throughout the seventeenth century, allowing the country to become, for a short time at least, a world power. The peat was used to fire the important industries that underpinned its economic development: saltworks, glassworks, breweries, dyers, potteries, and most lastingly the brickworks that helped create their beautiful cities, with their fine redbrick houses. Meanwhile, they used the new land that they had drained with lines of the famous Dutch windmills to grow cereals to feed the growing population. Nowadays this reclaimed land forms the heartland of the bulb fields that supply Holland's huge horticultural industry.

But the Dutch Golden Age was not destined to last long. By 1700 the low-peat reserves were largely exhausted, and it proved far more difficult and costly for the Dutch to access and transport the high-peat reserves that lay in the north of the country; building canals was not cost-effective. Meanwhile, the low-lying coast of Holland started to silt up, making access to its harbors increasingly difficult. Despite its lead in banking and commercial expertise, and its trading links with the Far East, the Netherlands lost its leading role to its rival across the North Sea—England.

While the Dutch had been lucky in the geography that had left them with a large deposit of easily exploited peat, England was even more fortunate. A combination of geography and geology gave it even larger reserves of energy that came in a far more concentrated form—coal. As we saw in the last chapter, per unit volume, coal contains around five times as much energy as wood and so fifty times that of peat. And Great Britain had huge reserves of coal, laid down during the Carboniferous period in several coalfields ranged roughly on a north–south axis, from the Welsh valleys in the south, to the central belt of Scotland in the North. The most fortunate aspect was that one of the largest fields, the Northumberland and Durham coalfield, had readily accessible coal reserves lying just below the surface, close to the North Sea coast and along the banks of three major rivers: the Tyne, the Wear, and the Tees. Miners could dig huge quantities of coal from the mines in this area, which were transported down to the rivers using wagons that rolled along wooden trackways—the forerunners of the railways—and loaded onto ships for onward transport. The greatest concentration of mines was around the shores of the Tyne, and Newcastle soon became a byword for the fuel, hence the expression "it's like taking coals to Newcastle." It grew rapidly to become the fourth-largest city in England by 1700, and energy-intensive industries such as salt making and lime burning quickly sprang up along the river and nearby coast. But the majority of the coal was transported south down the coast using increasingly large "colliers": broad-beamed and shallow-bottomed ships that were built in the area. Together with their sailors, these sturdy vessels formed a reserve of shipping and seamen for the Royal Navy, which was to later help Britain in its colonial enterprise; Captain Cook's famous ship *Endeavour*, which he used for his first voyage to Tahiti, New Zealand, and Australia, was a converted collier.

Much of the coal from Newcastle was finally shipped up the river Thames to power the rapidly expanding city of London. Coal use rose from around 150,000 tons per year in 1600 to almost 500,000 tons in 1700, supplying a population that grew from 200,000 to 575,000 in the same period. The environmental effects of coal burning were immediately noticed, and people were already complaining about the fumes it produced in the middle of the sixteenth century; Elizabeth I even brought in a bill forbidding coal fires. Nevertheless coal was so cheap that Londoners ignored all the environmental warnings and legislation. They used coal not only for heating, but also to fuel energy-intensive industries that they set up on the banks of the Thames: breweries, glassworks, saltworks, dyers, and metalworks. "Sea coal" also fired the brickworks that used the local London brick clay to make the multicolored bricks with which London was rebuilt after the Great Fire of 1666.

After the English economy was opened up along Dutch lines in the middle of the seventeenth century, following the social and political upheavals of the Republic and the Restoration of the Monarchy in 1660, the coal-powered industries expanded. With people and industry located together, London, dirty and smelly though it was, had all the advantages that enabled a new open mind-set to flourish. Intellectuals flocked to the city from the university towns of Oxford and Cambridge, meeting at the new English-language Gresham College, and in the newly opened coffeehouses of the capital. This brought together individuals with a bewilderingly wide range of interests and enthusiasms: the chemist Robert Boyle, who used his air pump to derive many of the laws of gases; the scientist and architect Christopher Wren, who rebuilt St Paul's Cathedral and the majority of the city's churches after the Great Fire; the doctor, statistician, and economist William Petty, who designed and tested a catamaran a hundred years before Captain Cook saw double canoes in Tahiti and Hawaii; and John Evelyn, who had interests in gardening, forestry, and engraving as well as air pollution. These natural philosophers rubbed shoulders with well-connected politicians and administrators, such as the diarist

Samuel Pepys, who was employed organizing the new Royal Navy. Together, they set up the most famous scientific society of all time, the Royal Society, and its first secretary, Henry Oldenburg, opened up scientific inquiry still further by instigating correspondence with scientists from around the Continent: people such as the Dutchmen Christiaan Huygens, the inventor of the pendulum clock; and Antonie van Leeuwenhoek, the early microscopist; and the German Gottfried Wilhelm Leibniz, the coinventor with Newton of calculus. The society was no collection of ivory-tower thinkers. For a start its royal patron Charles II was keen for it to help him on practical matters such as navigation. And its founders had all been inspired by the writings of the pioneer English scientist, Francis Bacon, who, in *The New Atlantis* of 1627, had advocated a public repository of knowledge, Solomon's House, which would be dedicated to "Knowledge of Causes and secret motions of things and the enlarging of the bounds of Human Empire to the effecting of all things possible."

Most important of all, though, to the success of the Royal Society and to its practical outlook was its curator of experiments, Robert Hooke, who was essentially the first professional scientist. A former assistant to Robert Boyle, he was paid to conduct all the experiments for the weekly meetings of the society and transcended all the divisions of science. A gifted all-round scientist, and part-time architect (he helped Christopher Wren rebuild the city of London), he mixed with the city's craftsmen and instrument makers and strongly influenced the society to undertake and publish useful applied research. He also promoted Joseph Moxon, the first tradesman to become a fellow, and helped him publish the first-ever set of DIY manuals, the *Mechanick Exercises*, which included instruction on smithing, joinery, wood turning, bricklaying, and making sundials. Even the gentlemanly Evelyn was moved to write one of the great bestsellers of the time. Although full of pretentious literary allusions and passages of whimsy, *Sylva* sought to bring together all that was known about the types of trees in the country and how they were grown and exploited. From the late seventeenth century onward, therefore, technological information became widely available, with rapid benefits to material progress.

Perhaps the most important invention to spring from this first flowering of applied science was the steam engine. The ever-expanding demand for coal had put the mining industry under increasing pressure. Miners were forced to dig ever-deeper underground to obtain the coal, and they soon ran into problems with flooding. The water levels in mines were traditionally lowered by cutting shafts or soughs that let water drain to outlets on the seashore or along the banks of rivers, but this technique could not, of course, drain mines below sea level. The colliers needed to pump out water actively, something that quickly grew beyond the capabilities of man- or horse-powered pumps. The principle of converting energy from coal into steam power was first demonstrated in the late seventeenth century by the French physicist and inventor Denis Papin. He had worked with Robert Boyle in London on an early pressure cooker and then went on to build a model engine in 1690. Papin's ideas were finally turned into a practical engine in the early eighteenth century by the Cornish engineer Thomas Newcomen. His atmospheric engine was powered by injecting cold water into a steam-filled cylinder, which condensed the steam and allowed the atmosphere to force a copper piston into it. The cylinder was then refilled with steam, pushing the piston gently back outward. This linear engine proved ideal to pump water out of coal mines, where its only disadvantage—high coal use—was no problem at all. Britain's coal supply had been secured, at least for the next half century, enabling industrialization to continue.

The heady early days of the Royal Society could not last. The inevitable process of specialization robbed it of some of its impetus, and Hooke was progressively marginalized by the aristocratic establishment. Under Isaac Newton's presidency the society became obsessed with the ideal of the "philosopher scientist" and progressively devalued the prestige of applied science, cutting British science off for a century or more from its important role in society.

Fortunately, by then applied science had taken off elsewhere in Europe, particularly in France, where the famous *Encyclopédie*, edited by Denis Diderot and Jean le Rond d'Alambert, stimulated a myriad of encyclopedias of technology that spread knowledge across the world. And the center of industry in Britain moved away from London and to those areas blessed with reserves of coal. In the eighteenth century, self-made men from all around the provinces sought to combine this new energy source with other geological riches of the island to develop and expand a wide range of industries. In Sunderland, they used coal brought down the river Wear to power the kilns of a new glass industry that used as its raw materials the local sand and Magnesian Limestone. Along the Yorkshire coast, they used coal to boil the vats in the alum works using chemicals from the local Jurassic shales. In Stoke, local coal from the Staffordshire coal seam powered the kilns of a new ceramics industry that used as its raw material the adjacent clay deposits. Staffordshire coal was also used to crystallize out the salt mined from beneath the

South Cheshire salt towns, Nantwich, Middlewich, and North-wich, and to heat the breweries of Burton on Trent. And the coal of the West Midlands was used to fire the hearths of the Birmingham jewelry and metal workshops. Britain was starting to become an urban nation, with expanding industrial towns, each powered by its own coalfield, and with intellectual societies of their own, such as the Lunar Society of the West Midlands and the leaders of the Scottish Enlightenment in Edinburgh.

The benefits of the new cheaper energy source had spread from London right across the country, had helped inspire new technologies, and had broken the power of craft guilds. The industrialists set up a world of manufacturing that produced goods that were cheap enough and attractive enough to persuade people to replace the perfectly good wooden ones with which they had formerly been perfectly satisfied. Cheap pottery replaced wooden bowls and plates, despite its fragility; pewter and glass tankards replaced wooden goblets. And the need to transport these fragile new goods smoothly and cheaply to market helped provide the impetus for capitalists to invest in a new transport system. In the second half of the eighteenth century canal mania was set off; new canals were built, crisscrossing the country to transport goods from the manufacturers to the customers in the towns and coal from the mines to the newly developing manufacturing districts.

There was just one missing piece in the energy jigsaw that prevented all industry from concentrating in the expanding towns and that limited Britain's industrial output; one by-product of wood, charcoal, was still needed to smelt iron. The iron industry in England was then run on a surprisingly small scale. Even as late as 1700, it produced only around twenty-five thousand tons of iron a year, roughly a thirtieth by weight of the amount of timber used in the country, and two hundred times less by volume. The problem was that the industry was limited by the supply and price of charcoal. Despite that huge areas of woodland in the Weald of southeast England had been placed under intensive coppice management to provide a large sustainable output of cheap wood, charcoal was still dearer in England than on the Continent. Dotted around the Kent and Sussex countryside some two hundred miles from the industrial heartland of the country, the ironworks were also poorly placed to innovate and collaborate with the rest of industry. English iron was becoming uncompetitive, and the country imported around twenty thousand tons a year of iron from Sweden and Russia, which had less pressure on their woodlands and purer iron ore, and which consequently had a competitive advantage.

The obvious solution to reduce the price of iron, localize production closer to the rest of industry, and increase output was to smelt iron not with charcoal, but with coal. After all, it was becoming much cheaper than charcoal, and was already almost pure carbon. Moreover lumps of coal are stronger than brittle lengths of charcoal, and so can be loaded into much larger blast furnaces without collapsing under the weight of the ore. However, ironmasters had always had problems smelting with coal. Its compacted nature gave it a lower surface area than charcoal, which is perfused with the lumens of its wood cells, reducing coal's rate of chemical reactions. Even more of a problem was that most coal contained impurities such as sulfur, which contaminated and weakened the iron it produced. Various techniques were developed around the world to overcome these difficulties, and iron was smelted with coal as early as the fifth century BC in Jiangsu, China. However, in Europe it was only in 1709 that Abraham Darby I, an ironmaster in Coalbrookdale, Shropshire, found a way to exploit the coal reserves that lay all around him. In the method he developed and patented, he simply heated local coal in large furnaces, a process similar to charcoal burning; this burned off the impurities and produced coke, a purer, stronger granular mass of carbon. He could then pile his coke in new larger and taller blast furnaces, which could burn hot enough to fully melt iron. His process formed a liquid metal that could be poured into molds, to produce what was known as pig iron.

Darby made a fortune, reheating the pig iron and pouring it into his patented sand molds to make cheap cast-iron pots and pans that undercut traditional copper and brass cookware. His new iron-making method was slow to catch on elsewhere, however, and little cast iron was produced in coke furnaces before 1750. The main reason for this was that the iron it produced tended to be brittle and hard to use. Darby had been fortunate that the coal from the East Shropshire coalfield was remarkably pure, only containing around 2 percent sulfur. Darby's method of coke burning could remove this, but not the higher levels of sulfur, up to 7 percent, found in other types of coal. Not until the 1750s did other ironmasters develop techniques that removed it effectively; they used steam-powered bellows to blow air onto the iron, increasing the temperature of the forge, and added limestone to react with and remove the sulfur.

Freed from the need to be located within vast woodlands, the iron industry quickly moved northward and westward to the coalfields of the industrial heartlands of Britain: to Shropshire, to North Wales, to Sheffield in Yorkshire, and to the Black Country near Birmingham. Many new industries were developed that exploited the benefits of cast iron. It was easy to mold into complex shapes, so it proved ideal to make all manner of decorative objects, fence work, garden furniture, and the like. And a new breed of agricultural engineers found it useful for making farm tools such as scythes and hoes. Most important were new plows with curved moldboards that could be pulled through the soil with less effort than traditional wooden plows and turn the soil over in a furrow, rather than just scratch its surface. James Small's Scots plow of 1784 was the first plow to be cast entirely in one piece of iron; it could be pulled by just two horses and operated by a single man, greatly increasing the productivity of the agricultural labor force. Finally, with its higher melting point, cast iron was better than other metals at withstanding heat, so it proved ideal to make pots, pans, fireplaces, and all manner of stoves. Iron troughs replaced lead ones throughout industry as wood-powered furnaces gave way to hotter coal-powered ones.

But cast iron was used in the largest quantities to make cannon, especially after another ironmaster from Coalbrookdale, John Wilkinson, developed the first precision cutting tool that could bore out a smooth barrel from a solid cylinder of metal, producing a safer, more accurate weapon. As the eighteenth century neared its end, the arms race between the navies of the European powers fanned the rise in the numbers, and size, of the ships, and the number of cannon with which they were armed. By the time of the Napoleonic Wars, the British fleet had some fourteen thousand cannon, ranging from the largest, 32-pounders (guns that fired thirty-two-pound cannonballs), each gun weighing over three tons, to smaller 16- and 8-pounders, and still smaller guns firing grapeshot and mortar shells. These guns alone would have used around twenty-five thousand tons of iron; this would have been a year's production in 1700.

John Wilkinson's new boring tool also helped develop another invention of the industrial revolution that enabled even more coal to be used: the steam engine. In 1765, James Watt had patented the invention that was to dramatically improve the efficiency of steam engines and turn Newcomen's atmospheric engine into a workable power plant. Watt devised a separate condenser to remove steam from the cylinder and create a vacuum to drag the piston down without actually cooling the cylinder and wasting heat. Watt's problem was that he could not make the cylinders of this engine perfect enough to be airtight; steam continually leaked out of them, wasting heat and power. The problem was quickly overcome by Wilkinson, who had ordered two steam engines from Watt. Wilkinson simply used his cannon-boring technique to cut a perfectly circular cylindrical hole through a solid iron block, to create the smooth inner surface of the cylinder, which allowed the piston to fit snugly inside the cylinder and move freely up and down. The major technical difficulty in making steam engines had been overcome.

It is hard to overestimate the importance of this development. Engines designed by Watt, built by the firm of Boulton and Watt, and with cylinders cut by Wilkinson were soon being shipped in their hundreds across Britain and the Continent. Forges and mills no longer needed to be built next to fast-flowing but temperamental streams. Instead of being powered by wooden waterwheels, whose power could be lost in dry weather, and which could be damaged in floods, they could use iron steam engines that could work twenty-four hours a day, whatever the weather. Instead of being dotted around the countryside next to streams, industrial plants could be sited together in larger and larger towns: in the iron-making Black Country of Wolverhampton and Birmingham; in the cotton towns of Lancashire; and in the wool towns of Yorkshire. With industry concentrated in this way, innovations could be passed on more quickly, industry could be made more efficient, and material progress could speed up—at the expense, it must be said, though, of the working and living conditions of the mill workers. The influence of iron making and the steam engine together transformed Britain's energy economy. From 3.5 million tons in 1700, Britain's output of coal rose to 6 million tons in 1750, and 17 million tons in 1800, producing twenty times the energy that wood had provided in 1650. By 1800 Britain had almost totally replaced wood as a source of fuel.

Elsewhere in the world, however, the influence of coal was far less than in Britain, and the Age of Wood persisted. Other European countries had far smaller coal reserves, and the coal they had was usually far less accessible, so continental countries depended far more on wood for fuel. The governments in France and Germany became increasingly worried about wood supplies, and unlike in Britain, there was pressure to nationalize the forests and to start to run them on more "scientific" lines. They built up a bureaucracy and infrastructure to manage forests, in effect farming their trees in a far more systematic way. European rulers also hastened to maintain their wood supply by constructing large-scale engineering works to transport logs down rivers to the cities. Wood suppliers became larger and larger commercial enterprises, particularly in the Black Forest, and were controlled by moneymen, not the local inhabitants. They built bigger chutes to transport logs down to the major rivers, and far larger log rafts. By the end of the eighteenth century, Dutch rafts on the Rhine were composed of several sections and resembled floating villages, with rudders up to forty-five feet long, operated by seven men. The main section could be four hundred yards long and ninety yards wide and was manned by several hundred oarsmen, who were housed in huts on the raft, along with several classes of passengers. The navigator sat atop a viewing platform raised some ten feet above the deck. Given the lower energy density of wood, it's no surprise that on the Continent, canals were built far wider than their counterparts in England, and the goods were transported by huge barges, rather than charming narrow boats.

On the other side of the supply-and-demand equation, continental scientists and inventors became obsessed with improving the efficiency with which they burned wood. They developed new fuel-efficient kilns, and people swapped open fires for iron wood-burning stoves that retained far more of the heat within the house; efficiency rose from 10–20 percent for open fires, to 40–60 percent. This change also had a major impact on continental forests. Rather than burning slender poles, wood-burning stoves work best when they are fed with quite large pieces of wood, so during the eighteenth century foresters in France and Germany increased the length of their coppice rotations to fifty and eventually to eighty years, effectively converting coppice to plantation high forest. And since softwoods could grow faster in these conditions and floated better than hardwoods, the foresters replanted broad-leaved woodland with conifers.

The fast-growing colonies of North America, and the newly liberated American nation, meanwhile, had no problems at all with energy supply. The eastern states were covered with apparently limitless forests, and the developing towns and cities were situated on the coast or along the banks of large rivers or lakes, giving them easy access to logs or sawn timber floated down from the interior. They rapidly threw up settlements that were essentially timber versions of European ones, and built grand timber mansions in the graceful "colonial" style. And rather than digging for coal, American industrialists developed new techniques to improve the quality of charcoal, which enabled them to carry on producing competitively priced iron. Instead of burning wood in small stacks to make charcoal, they heated it indirectly in huge retorts, which not only produced a higher yield of charcoal, but damaged it less so that the pieces of charcoal were far stronger. The process also released a host of useful chemicals. The tar, methanol, and acetone produced as by-products were sold off, which made the industry more competitive. With their stronger charcoal, American iron makers were able to smelt iron in furnaces that were almost as tall as European coke furnaces; not until 1850 did coke-based production grow larger than that of charcoal, and not until the end of the 1920s did charcoal-smelted iron production effectively cease. And instead of using coal, Americans modified the design of their steam engines to burn wood. America, more than any other industrial nation, stayed rooted within the Age of Wood right up to the start of the twentieth century.

CHAPTER 12

Wood in the Nineteenth Century

For a period during which industry and science had been expanding so dramatically, and which had seen the rise of great European empires and the American and French revolutions, the events of the eighteenth century had surprisingly little effect on the infrastructure. Modern moviegoers are often hard-pressed to spot the differences in the backdrops of films set early in the century, *The Favourite* or *Gulliver's Travels*, for instance, and *Amadeus* or *Pride and Prejudice*, from the end, and for good reason. The buildings, designed in the neoclassical style, were much the same in proportion and all built on the same human scale. Even the new factories built at the end of the eighteenth century were of modest size, just a few stories high, and from the outside they looked just like the warehouses that had been built at the century's start. People still traveled around towns and across the country in horse-drawn wooden carriages, and around the world in wooden ships. The main differences that could give viewers a clue to the period were seen indoors: the change in furniture, from heavy oak to elegant mahogany; the increase in the amount of china and other ornaments on display; and the change in fashions from delicate oriental silk dresses to simpler European cotton gowns.

The reason for this apparent paradox was that even though the British had broken free from the energy constraints imposed by using firewood and charcoal, their technology and engineering was still limited by its reliance on timber. The time needed to grow and fell trees, cut and season timber, and to ultimately shape it into useful products limited productivity, while the engineering disadvantages of wood limited the size and strength of the structures into which it could be made. As we have seen, timber is prone to splitting, difficult to join, vulnerable to warping and rotting in the open air, and flammable. Carriages were heavy and uncomfortable, and wooden ships remained small, cramped, and leaky, and this conspired to limit travel. At the end of the eighteenth century it still took three to four days to travel the four hundred miles from London to Edinburgh. Sailing times barely improved either. It took Benjamin Franklin some seven weeks to travel across the Atlantic on his first sailing in 1724; his last voyage in 1786 took exactly the same time. Meanwhile, bridges with a span of more than a hundred feet were rare, buildings were limited to six stories in height and covered spaces to eighty feet in width, and the wooden machinery that filled the mills was heavy, crude, and clumsy.

Considering the great increase in iron production in the second half of the eighteenth century, one might have expected the situation to have improved. Unfortunately, however, the cast iron that the ironmasters produced proved to be no substitute for timber. Cast iron contains high levels of carbon, usually around 4 percent, and unlike the bar iron produced by blacksmiths, it does not contain any slag fibers that would have reinforced it; this meant it was brittle. Though it could withstand large compressive forces, it broke easily and unpredictably when it was stretched, as cracks ran readily through it. Just like stone, it could not be used to make chains that could withstand being stretched or beams that could withstand being bent, and it could not withstand impacts. As an engineering material, it could only safely be used to replace masonry, not wood.

This is admirably illustrated by what is perhaps the most celebrated structure of the early industrial revolution, the Iron Bridge over the River Severn at Coalbrookdale. Built from seventeen hundred cast-iron components in 1784 by Abraham Darby III, the grandson of the pioneer of cast iron, this impressive structure had a span of 120 feet. However, though the bridge was put together like a wooden one, each of the pieces being joined together using dovetail or mortise-and-tenon joints, it was an arch structure, and the iron was loaded purely in compression. Its 380-ton filigree of iron simply replaced what would have been a much heavier stone arch. Following the success of the Iron Bridge, arches of cast iron were used to build many other smaller road bridges, and the great engineer Thomas Telford also used cast iron in the construction of the spectacular Pontcysyllte Aqueduct over the river Dee near Llangollen, north Wales. He held the water in cast-iron troughs and supported these between the stone pillars of the aqueduct with cast-iron arched ribs.

The success of cast-iron arches tempted early-nineteenth-century engineers to use cast iron more extensively to build railways and to replace structural timbers in their mills. However, this caused a series of disasters. The floors of the first iron-framed mills were all too apt to collapse under the bending loads set up by the heavy machinery. Cast-iron rails laid down for the expanding railway system continually broke under the dynamic loading from heavy trains. And cast-iron truss bridges often failed catastrophically. Robert Stephenson's Dee Bridge, for instance, collapsed in 1847, just as a train was crossing it, killing five people. By the middle of the nineteenth century, it was clear that cast iron should only be used in structural elements that were loaded purely in compression. So in what we usually think of as archetypal "cast-iron" buildings from the middle of the nineteenth century—railway stations and greenhouses—the only parts of the structure that were actually built of cast iron were the pillars.

The key to the ability of nineteenth-century engineers to build a new industrial world, to construct bridges of record length, buildings of unprecedented size and novel design, and gigantic ships was the invention of a rather different material: wrought iron. Ten times as stiff as wood, up to three times as strong in tension, and ten times as tough, wrought iron was the first material with mechanical properties superior to those of timber that could be made in large quantities and in large pieces. Invented in 1783–84 by an ironmaster from Lancashire, Henry Cort, wrought iron resembled traditional rod iron, but could be produced around fifty times faster. Cort's patented "puddling" technique involved melting a lump of cast iron in a furnace, so that it formed a puddle of molten iron on the floor. The puddler then stirred the metal with a long iron rod through a hole in the side of the furnace, causing the carbon in the metal to react with iron oxide that had been strewn on the floor of the furnace and bubble off as carbon monoxide. As the carbon level of the metal fell, its melting point rose, and it started to solidify, and the puddler would continue to roll and fold the piece of metal, incorporating fibers of the slag that had also been placed in the furnace. Finally the puddler would lift the roll of metal out of the furnace and it would be shaped and strengthened by being squeezed through a series of rollers, to convert it into a long metal rod or flat metal sheet. The slag made the wrought iron fibrous, toughening it and protecting it from rust, just as in traditional bar iron, but a single puddler could make 220 pounds of wrought iron in a single day. It was soon being made in huge quantities and opened up a new world of possibilities.

Engineers quickly started to exploit the high tensile strength of wrought iron, making it into chains to support a new type of bridge that would soon span gaps greater than any masonry or timber structure: the suspension bridge. By 1810 the Irish-born engineer James Finley had built a suspension bridge with a 240-foot span over the Merrimack River at Newburyport, Massachusetts, but bridges rapidly got bigger. Samuel Brown, a former captain in the Royal Navy, used his patented wrought-iron chains to support the deck of his successful Union Chain Bridge across the river Tweed. With a span of 449 feet it was the longest bridge in the world when it opened in 1820 and today still links the two countries of England and Scotland. This bridge was soon outdone, though, by Thomas Telford's Menai Suspension Bridge, which carried the route of the London to Holyhead turnpike road over the strait between Wales and Anglesey. Completed in 1826, it has a span of 577 feet, held up by just 123 tons of wrought iron, under a third of the amount of iron needed for the much-shorter Iron Bridge of fifty years before. By 1864, the world's longest bridge was Isambard Kingdom Brunel's Clifton Suspension Bridge over the river Avon at Bristol, which had a span of a hitherto unimaginable 702 feet.

But the railways proved to be a much greater consumer of wrought iron. The new locomotives needed to be both powerful enough to pull their carriages, and small and light enough to be mobile, so they had to operate at high steam pressures. To resist the stresses that the high-pressure steam set up in the boilers, locomotive designers constructed them out of wrought-iron plates that they riveted together to form pressure-tight cylinders. Even with the advantages of relatively lightweight engines, though, the railway engineers still had problems with their tracks. The wooden rails that had been perfectly adequate for the old wagonways were too flexible and wore out rapidly under the iron wheels, while cast-iron rails were too brittle and continually snapped under the dynamic loads produced as the trains passed over them. Engineers were forced to use wrought-iron rails, despite their higher cost, and ironmasters soon developed new methods to roll lengths of rail into the correct shape, which the railwaymen then laid out on stone sleepers.

Wrought iron also increasingly became the material of choice to build the large numbers of bridges the railways needed to cross rivers and valleys. Engineers soon found that suspension bridges were just too flexible: a heavy train crossing a suspension bridge would make it sway alarmingly and damage the trackway. Most railway bridges were therefore made using the principles of the beam, with the wrought iron resisting the forces of tension and compression set up when it is bent. One of the first great beam bridges was Robert Stephenson's 1846 Britannia Bridge over the Menai Strait, which he built just a few miles away from Telford's suspension bridge. Stephenson's solution to bridging a large span was to construct two wrought-iron tubes of rectangular cross section and run the trains through them. Though the bridge was a success, Stephenson quickly realized that a closed tube was unneeded. In his later bridges he replaced wrought-iron plate with a framework of crisscrossing struts, producing the first modern trusses such as the High Level Bridge across the river Tyne, completed in 1849. Soon, engineers were building huge numbers of truss bridges, long and short, all over the world, their massive lattices and crisscross patterns of struts being almost a signature of nineteenth-century engineering.

Wrought iron also changed the way in which buildings were constructed, allowing new types of structure to be built that would transform the urban environment. Following the destruction by fire of the Belper North Mill, Derbyshire, the industrialist William Strutt replaced it with one whose upper floors were supported by horizontal iron beams, the model for all subsequent European textile mills that were constructed around a complete wrought-iron frame. These new composite buildings could be built with their pillars much farther apart, providing open factory floors, and could be built several stories high; they became the prototypes for the steel skyscrapers of the twentieth century. The strength of the frame also meant that the masonry walls acted now just as a curtain, to keep out the wind and rain. They could therefore be replaced by another new wonder material—sheet glass—to build an entirely new form of building, the greenhouse. Architects and gardeners-turned-architects such as Joseph Paxton designed larger and larger greenhouses to hold the horticultural spoils of Britain's empire: palms, tree ferns, and giant water lilies. Their roofs were supported by wrought-iron girders, which were themselves held up on cast-iron pillars. It was Paxton who designed the greatest prefab of all time, the Crystal Palace, which was erected in Hyde Park in just eight months to house the Great Exhibition of 1851. Enclosing almost a million square feet of space and rising to a height of 168 feet, it was a triumph of engineering in wrought iron—though it must be said that it also incorporated a large amount of wood. The glass panes were held within Paxton's patented wooden glazing frames with their integrated guttering system, while the highest and grandest part of the structure, the great transept, was supported by sixteen laminated wooden arches, each seventy-two feet in diameter.

The same techniques were also used to build the structures that more than any other became the cathedrals of the age, the great engine sheds that sheltered the platforms of the new railway stations that were being built across the world. None is more spectacular than the great St Pancras arch in London, terminus of the old Midland Railway and now converted into the British base of Eurostar. A single open space 690 feet long and 240 feet wide, the station was the largest unsupported covered open space in the world when it was opened in 1868. Each of the twenty-nine arches spans a distance three times as wide as the Palazzo della Ragione in Padua, the nearest wooden equivalent.

But of all structures, wrought iron transformed the building of ships the most. We saw in chapter 10 that not until the early nineteenth century did shipbuilders first start to understand the need for diagonal bracing in wooden ships. Engineers soon realized that an even better, and far simpler, method of producing a rigid, watertight hull was to rivet together sheets of wrought iron into a tubular hull, which could be stiffened with internal bulkheads and stringers. Without the complexities of internal frames and external planking, ships could be built larger, more quickly, and more cheaply. Britain's most charismatic Victorian engineer, Isambard Kingdom Brunel, for instance, designed and built three ships in his career, each larger than its predecessor, and unprecedented in size. His first, the paddle steamer the SS *Great Western*, built in oak and launched in 1838, was

235 feet long and had a displacement of twenty-three hundred tons. Though the *Great Western* was then the largest passenger ship in the world, it was soon dwarfed by his next ship, SS *Great Britain*, the first iron ship to be driven by screw propellers. Launched in 1843, it was 322 feet long with a displacement of thirty-seven hundred tons. Brunel's final ship, the SS *Great Eastern*, launched in 1859, was 692 feet long with a displacement of nineteen thousand tons.

The new iron ships had another advantage; the toughness of wrought iron gave them far more protection from cannon balls and shells. By the 1850s it was becoming apparent that an ironclad vessel could destroy any wooden ship, while being invulnerable itself, something that was finally conclusively demonstrated at the Battle of Hampton Roads in 1862; the Confederate ironclad CSS *Virginia* destroyed two wooden Union ships, but was powerless to defeat the Union's own ironclad USS *Monitor*. In Europe, the French ironclad *Gloire* and Britain's HMS *Warrior*, the world's first completely iron-hulled warship, sparked off an arms race between the two countries that was to last until the end of the century.

The wrought-iron age climaxed in two iconic structures that look superficially totally different, yet which were engineered by the same man. To the casual onlooker, the Statue of Liberty resembles a conventional, if enormous, bronze sculpture. It looks solid, but anyone who has climbed it knows that what you see from the outside is just a thin skin of copper. The secret of the statue's stability is its internal structure, a massive space frame of wrought-iron rods, a masterpiece of French engineering, anchored at just four points at its base to the American-built stone podium. The frame was designed by none other than Alexandre Gustave Eiffel and looks like an asymmetrical version of his most famous structure, the Eiffel Tower. Just as the Statue of Liberty is iconic for the USA, so the Eiffel Tower is for the French. Rising 1,063 feet in height and constructed from ten thousand tons of wrought iron, the tower was completed in 1889 and for forty years was the tallest man-made structure in the world.

The wrought-iron bridges, railway stations, and ships of the nineteenth century are so spectacular that it is tempting to say that at this point iron eclipsed timber. But as we have seen earlier in this book, people have time and again exploited novel materials to improve the ways in which we use wood; wrought iron was no exception. One of the first examples is the way in which it helped speed up the manufacture of wooden items that were needed by every sailing vessel—ships' blocks. Just a stone's throw from HMS *Victory*, Nelson's flagship at the Battle of Trafalgar, and the Royal Navy's first all-iron ship, HMS *Warrior* in the old naval dockyard in Portsmouth, England, there are some old disused sheds. The Block Mills don't look like much, but are historically even more important than the magnificent ships, marking not only an early example of how wrought iron gave new life to wood but also a major stepping stone in the industrial revolution.

Until the end of the eighteenth century, England had lagged behind Europe and the United States in its use of sawmills. English carpenters and shipwrights preferred to have their timbers cut by hand, using two-man saws that produced a beautiful cut surface. The system was ideal to shape one-off items, such as a bespoke keel. However, in 1796 the new inspector general of naval works, Samuel Bentham, a naval architect and talented engineer, realized the importance of rearmament in the face of the threat from revolutionary France and recognized that hand sawing was just too slow. He oversaw the construction of a sawmill in the dockyard, powered by a Boulton and Watt steam engine; just a few men could now cut the large numbers of identical small timbers needed to build ships. But he also realized that he had a supply problem with one standardized component that all sailing ships of the time needed in large numbers— ship's blocks or pulleys.

A single three-masted ship of the line could have over a thousand pulleys in the complicated rigging, which were essential to enable the crew to set the sails. The navy employed hundreds of men in separate shipyards around the country hand-making these items, but this was expensive, and the supply chain was unreliable. Being an enthusiastic engineer himself, Bentham designed some machines to speed up the job. However, before he could put his plans into operation, he met another engineer whose ideas Bentham quickly realized were superior, Marc Isambard Brunel, father of the Isambard Kingdom Brunel we met earlier in this chapter. A refugee from revolutionary France, Brunel had recently returned to Europe from New York. He realized that for such small objects as blocks, he could treat wood just like any other material, one that could be machined in precise industrial operations to make and assemble large numbers of identical items. To make each block he therefore designed machines that could cut and shape each of the three main elements of the block so accurately that they would always fit together: the main body of the block, called the shell; the rotating pulley itself, called the sheave; and the axle for the pulley, called the pin. For instance, to make the shell, the first machine used a circular saw to cut out a rectangular block of wood. The second drilled a hole for the pin, and two others at right angles to mark the ends of the groove in which the sheave would rotate. This machine also indented locating points so that the shell would be held correctly in the subsequent machines. The third machine used mortising chisels to cut the rest of the groove, and the fourth cut the corners off the block to give it its tapered shape.

The Admiralty were enthusiastic about the proposal and organized for the machines to be tested, employing a young engineer, Henry Maudslay, who had already built up a reputation for constructing precision security door locks. The collaboration between the two engineers proved fruitful, and the final machines, made from wrought iron, worked perfectly. In all, forty-three machines, carrying out twenty-three processes, produced three sizes of blocks. Not only were these among the first machine tools, but they were arranged in order, making them into perhaps the world's first production line. Powered from the steam engine by a series of pulleys, just like the spinning machines and looms in the textile mills of the time, the factory was a great success; the 130,000 blocks that were needed each year were produced by just ten men. So perfect were the machines that they continued churning out blocks for over 150 years; production only ceased in 1964, with the final demise of naval sailing ships.

Brunel's block factory was an example of how machines could be used to manufacture wooden objects that had until then been made by hand. Soon, however, engineers and builders also developed novel structures that combined wood with wrought iron. Using wood in this fresh new way enabled them to make even more use of it. They exploited the advantages that wood had always had—its ability to withstand bending, its lightness, stiffness, and toughness—and found new ways to minimize its disadvantages—its variability, tendency to split, and the difficulty of making carpentry joints. And here we shift our geographical focus, for most of the novel ways of using wood were invented not in Britain, which was at that time obsessed by iron, but in the countries that were blessed by huge forests and which had remained firmly in the Age of Wood, in North America and Scandinavia.

Wrought iron proved to be particularly helpful in replacing weak carpentry joints, such as mortise and tenons and dovetails, with stronger metal joints between the structural elements. Perhaps the earliest example of this trend is seen in king-post roof trusses, in which a wrought-iron sling that was bolted to the king post was wrapped below the tie beam to hold it up, creating an efficient triangulated structure. American engineers soon realized the advantages of drilling through the wooden elements in their structures and joining them with wrought-iron rods. As early as 1812, for instance, Louis Wernwag used this technique to build his famous Colossus Bridge over the Schuylkill River near Philadelphia. The bridge, a shallow arch, was at 340 feet the longest wooden bridge in the world; the main wooden elements of the trusses that supported it along its length were connected by iron rods and made apparently "without a single mortise and tenon joint." This was an important consideration for a country of immigrants, where labor was expensive and skilled carpenters thin on the ground.

Unlike in Europe, where the railways were built with iron and stone, Americans built their railroads for the most part with timber. They used steam engines with wrought-iron boilers, of course, but they mounted the rails on wooden sleepers, while they built wooden railway carriages whose timber frames were joined together using iron rods and bolts. They even experimented with wooden rails. Logging trains were often run along pole roads—essentially just logs laid parallel on the ground—staying on the track by using wheels with cup-shaped rims like huge pulleys. The temporary tracks cost only $75–$250 per mile, though they soon rotted away. And early mainlines were often built with strap rails, thin strips of wrought iron pinned to a wooden rail. But the glory of the American railways was the spectacular wooden trestle bridges they built to cross gorges, replacing the iron trusses and soil embankments beloved of European engineers with cheaper, quicker-to-build frameworks of timber beams and iron joints. The result was an exceptionally cheap railway—American railroads typically cost $20,000–$30,000 per mile, less than a sixth of the $180,000 a typical European railway cost. It was this, more than any other factor that enabled the Americans to build their long-distance railroads so early, from the mid-1830s onward. The new lines opened up the West Coast to new colonists and proved a far quicker and safer alternative to the old wagon trains of the Oregon Trail. It was left to a later generation to upgrade the tracks with full wrought-iron rails.

Novel wooden structures also helped speed up the economic and political development of the West Coast. The discovery in 1859 of vast reserves of silver ore in the Nevada mountains led to a silver rush that was just as important as the Californian Gold Rush of the previous decade. One particularly large deposit was the Comstock Lode, which contained vast quantities of soft ore that could be readily extracted with shovels. The problem was that the surface drifts were soon exhausted and the difficulty was to extract ore safely while avoiding deadly cave-ins. In coal mines wooden pit props were traditionally used to hold up the workings through the narrow coal seams, and Britain's mines relied on the import from Canada of huge numbers of six- to eight-foot-long spruce logs. Using simple props was simply not possible in the Comstock Lode where the deposits could be hundreds of feet thick. It was a German mining engineer, Philip Deidesheimer, who came up with the solution: square set timbering. As the miners removed ore, they replaced the space with six-foot cubic latticeworks of timber, each one being linked together with wrought-iron rods to form a gigantic framework that propped up the entire lode and which could be backfilled with waste. The Comstock Lode made vast fortunes, including that of the influential Hearst family, payed for the expansion of San Francisco, and prompted the incorporation of the state of Nevada into the Union.

Americans were also quick to recognize the advantages of using an even simpler and humbler attachment system—nails. Nails had long been handmade in what amounted to a craft industry. The town of Bromsgrove near Birmingham, England, for instance, housed nine hundred nail makers at its peak. However, handmade nails were expensive, and in Europe no carpenter worth his mettle would deign to use such crude devices. In America the situation was quite different. There were no craft guilds, but large numbers of settlers needed houses and huge amounts of wood they could build them with. The switch to a new way of building houses was facilitated by the invention of machine-cut nails, the first machines to cut them from wrought-iron sheet being patented in the USA by Jacob Perkins of Massachusetts in 1795. Unlike modern wire nails, these "cut nails" were square in cross section, and since they cut through the wood as they were hammered into it, rather than splitting the cell walls apart, as modern circular nails do, they were much better at holding on. The nails soon started to be produced in bulk, and their price fell dramatically.

By 1830, American builders were starting to exploit the wood-holding capabilities of nails to erect cheap, mass-produced houses, making use of increasingly sophisticated sawmills that were equipped with precision steel saws. Rather than just cut logs into thick, heavy beams, the new mills were able to slice them into uniform thin planks and narrow studs. The builders could then nail these uniform elements together to make the frames of houses, in a technique dubbed balloon framing because of its lightness and apparent fragility. The main structure of the house was built with long, narrow vertical studs, the famous two-by-fours, which stretched all the way from the floor sill plate to the rafter plate in the roof, and which were nailed to horizontal lintels and ribbons. The walls could be fully assembled on the ground, then simply lifted into place and nailed together. The only grooves and joints were in the floor joists, which were let into and nailed to the studs and rested on the ribbons. Finally the whole structure was sheathed on the outside with planks, and on the inside with boards, and insulation could be placed in between the two skins to keep the house warm in winter and cool in summer. It is hard to overemphasize the importance of balloon framing in allowing American settlers to be cheaply and decently housed; despite looking flimsy, balloon-framed houses proved surprisingly robust. Even today most Americans live in wood-framed houses, though since 1940 most have been built using a rather different system, platform framing. In this new technique the vertical studs only extend for a single floor. Each story of the house is built as its own box, in much the same way as the half-timber houses of medieval Europe, and the boxes are then laid one on top of the other. To a Briton such as myself, used to brick and stone houses, it is disconcerting to see how thin the walls of American houses are. It certainly means less space for knickknacks and pot plants on the windowsills, but I have to admit that the system is far more economical of materials, and the houses are just as cozy and warm up far faster.

If nails proved useful for building houses, another wrought-iron invention, the machine-made wood screw, made all sorts of structures, from furniture to fencing, quicker and cheaper to make. The first screws were made in the Middle Ages, being used to attach the leather to the wooden frames of bellows and to join metal plates together in suits of armor. But they were rarely used in woodworking, largely because they were so difficult to make and consequently expensive. Each screw had to have its thread hand-filed, a fiddly and lengthy process, which produced crude and variable results. The first machine-made screws were developed in 1760 by two brothers, Job and William Wyatt from Burton-on-Trent, England. Their technique successfully made reliable screws, but as these were cylindrical in shape, more like bolts, the carpenter had to drill a hole in the wood before they could bite into it. Only in the 1840s did a succession of American inventors, Cullen Whipple, Thomas J. Sloan, and Charles D. Rodgers, patent ways of making workable pointed screws, and from then on American companies dominated the screw-making business. At last, even woodworking amateurs could make reasonably strong joints and construct useful structures at home. Nowadays screws have been joined by a wide range of ingenious jointing mechanisms that have been designed to hold together the different elements of flat-packed furniture. Millions of people across the globe have been introduced to the delights of woodworking using such methods as they assembled their latest IKEA purchases.

But if wisdom ranks above wealth, as the motto of my old school claims, one nineteenth-century development must be more important than any bridge or house: the ability to make paper from wood pulp. Paper is composed of a random felt-work of fibers, which is pressed firmly together to produce a solid surface, and which can then be covered with a waterproof coat of china clay to prevent ink from running across it. Early paper was made with the bast fibers, or phloem, of trees, but in Europe it was typically made using the long thin cells that compose the fibers of flax or cotton. These are made up almost entirely of cellulose within a hemicellulose matrix. Both flax and cotton are mainly used to make fabric, so historically the paper industry relied on the recycling of rags for its raw material. They were collected by rag-and-bone men, who then sold the material to rag merchants or directly to rag-paper makers. This arrangement worked well when literacy was limited and when news was highly censored. However, as education spread in the seventeenth and eighteenth centuries, and a new spirit of enterprise saw newspapers spring up, the supply of rags started to become strained, and people began to look for alternative materials. As early as the beginning of the eighteenth century, the French scientist René de Réaumur suggested that paper could be made from wood; he had observed wasps chewing up wood from a post and reasoned that they must make their paper nests by mixing it with bodily secretions. However, there are problems with using wood as a source of paper fibers. Unlike cotton and flax fibers, wood cells contain lignin as well as cellulose and hemicellulose. As we saw earlier in the book, this stiffens the cell walls, making them far harder and more difficult to break up and separate. Unfortunately, lignin also breaks down in light, so wood-based paper tends to turn yellow over time.

The first of the problems was overcome in Germany in 1840 by Friedrich Gottlob Keller, who developed a machine to hold sticks of wood against a grindstone lubricated with water. The resulting wood pulp started to be widely produced around the world from about 1870, but it made paper that was rough and quickly discolored, suitable only for ephemeral text such as newspapers. Despite its poor quality, machine-made wood pulp revolutionized journalism, as the cost of pulp fell dramatically; in the USA it fell from around thirteen or fourteen cents per pound to around two to three cents in the 1880s. The price drop saw a dramatic increase in the size of newspapers, and their circulation. For instance, the *New York World* fell in price from five to two cents between 1863 and 1882, despite doubling in size, and its circulation quickly rose to twenty-five thousand. Between 1880 and 1890 the amount of newsprint used by American newspapers increased sixfold, from just over 100 million pounds annually to almost 700 million. The growth of sensational stories and salacious gossip that filled the new column inches spread alarm among American commentators, who put it down to moral decline rather than to market forces. The approach was known in the United States as yellow journalism, possibly in reference to the color that the cheap newsprint quickly became, and grew highly influential. It has even been suggested that the sensational coverage of the sinking of the USS *Maine* by Joseph Pulitzer's *New York World* and William Randolph Hearst's *New York Journal* helped force the United States to start the Spanish-American War of 1898.

But the growth of any new medium always stimulates scare-stories from the rich and powerful who are invariably keen to maintain the status quo. The expanding newspapers, transported across the country by the new railways, brought information ever faster to the already literate American population. And in Europe they helped lead to a rise in literacy, especially among women, and fed the growing demand for the expansion of the franchise and the birth of the suffragette movement. And it was not only in newspapers where cheap paper made from wood pulp had an impact. Two chemical processes were finally invented that allowed wood pulp to be turned into paper suitable for turning into books such as the one you are now reading. The sulfite process, which removed lignin from the pulp using sulfurous acids, was developed in the 1850s and 1860s and had by 1900 become the dominant method of pulping wood. And this process was itself superseded by Carl Dahl's sulfate or kraft process, which used sodium sulfide and sodium hydroxide to break down the bonds that link the lignin to the cellulose. With cheaper, better-quality paper bringing down the price of books, writers were now able to cater to a much broader reading public. Rising above the sensationalist "penny dreadfuls" of the mid-nineteenth century, which were printed on cheap straw paper or newsprint, whole new genres of literature printed on better-quality paper emerged. Novels such as Mark Twain's *The Adventures of Huckleberry Finn* and Robert Tressell's *The Ragged-Trousered Philanthropists* could grapple realistically with social issues; in America, dime novelists introduced the public to the Western, building a heroic history for the new country; in Europe, Jules Verne and H. G. Wells could examine the consequences of progress with science fiction books such as *Twenty Thousand Leagues Under the Sea* and *The Time Machine*; the first detective books appeared, such as Arthur Conan Doyle's Sherlock Holmes stories; the first thrillers such as Erskine Childers's *The Riddle of the Sands* emerged; and the first spy books, such as Joseph Conrad's *The Secret Agent* were published; and at last books could just revel in the comedy of everyday life, from George and Weedon Grossmith's *The Diary of a Nobody* to Jerome K. Jerome's *Three Men in a Boat*.

By the end of the nineteenth century, wood's importance was hardly diminished. Wood pulp paper had helped transform the mind-set of people across the world and timber was still widely used. But it was in the New World where timber had the greatest physical and economic impact. The world order still seemed to be dominated by Britain and the other European powers, who were confident that they would maintain their position. But the United States had started to use its vast reserves of timber to build a unified nation and to catch up economically with Europe. The rapid logging of the Great Lakes area supplied the timber needed to construct the infrastructure of a modern state—one that looked like that of Europe, only made of wood. With its wooden railways, wooden houses, wooden factories, and even wooden roads, all of which could be constructed swiftly and cheaply, it even started to overtake Europe.

But there were downsides with this rush for growth. Many of the lightweight structures proved ephemeral, all too prone to rotting away or to be destroyed by fires such as the one that burned down San Francisco after the earthquake of 1906. And the rapid logging cleared great swathes of the forest. I remember visiting the last remaining twenty acres of white pine forest in Lower Michigan at the White Pine Park, Kent County: all that remained of the vast tracts of conifer forest that had once cloaked the peninsula. On the same trip I stayed for a night at a historic wooden house in the former logging town of Muskegon. It had a great ballroom with a sprung wooden floor and was very charming, but it was virtually the only building remaining from Muskegon's heyday. The town itself seemed to have little purpose, and there was little left to remind people of their history. This was unlike Europe, where even in post-industrial towns many old stone and brick buildings remain; the houses in the streets around my home in Manchester, for instance, were all older than the Muskegon bed-and-breakfast. Even in Latin American countries, many fine old stone buildings testify to their long colonial history. As we shall see, it was the new materials of the twentieth century that finally helped the United States build a more permanent infrastructure. And despite yet more scientific innovation, we shall see that wood continued to be important throughout the twentieth century and not only in the United States but all around the world.

CHAPTER 13

Wood in the Modern World

If you want to get a feel for the nineteenth-century industrial world that was created with coal and wrought iron, there is nowhere better than Manchester, England, the world's first industrial city. Its Science and Industry Museum re-creates cotton mills, houses massive working steam engines, and takes people on rides on replica steam trains from the world's first railway station. And all around the center of the city massive iron-framed mills and warehouses still dominate the streets, with their soot-blackened brick walls and towering chimneys. Together with the great iron railway stations, truss bridges, and brick viaducts, they make the visitor feel just like one of the characters in L. S. Lowry's paintings: mere matchstick men and women. Yet today these nineteenth-century cities seem positively archaic and Lilliputian compared with the downtown areas of the great twentieth-century cities of North America, the Middle East, and Asia. Here, gargantuan towers shimmer in the sunlight, planes fly overhead, and the sheer scale makes the people who crowd their sidewalks and the cars that choke their streets seem so small as to be almost invisible. We owe these changes to the rise of a whole new suite of "modern" materials: steel, concrete, and plastic, and to a new source of energy, oil. It seems to be a world in which wood has no place, yet once again we will find that this new technology has enabled us to use even more of it.

Materials technology did not stop with the invention of wrought iron; toward the end of the nineteenth century, wrought iron itself started to be replaced by two new engineering materials: steel and concrete. Metallurgists had long known about the advantages of purifying iron by removing the carbon and slag impurities. The pure metal could be hardened and strengthened by adding small, measured quantities of carbon or other alloy metals, making steel that was twice as strong as wrought iron. Benjamin Huntsman of Sheffield invented crucible steel as early as 1740, and Sheffield steelmakers used it to make superior tools and cutlery. However, crucible steel was expensive since it could only be made in small batches of around thirty pounds. Only in the latter half of the nineteenth century did the Bessemer converter and open-hearth furnace enable large quantities of high-quality steel to be produced that was consistently stronger and cheaper than wrought iron. Steel started to replace wrought iron from the 1880s onward. The Brooklyn Bridge, designed by John Augustus Roebling, was the first suspension bridge to be held up using steel wires rather than wrought-iron chains. With a span of 1,595 feet, it was 50 percent longer than anything built before. Another early triumph for steel was the Forth Rail Bridge in Scotland, the first modern cantilever bridge, which was completed in 1889.

Steel frames soon became the norm for tall buildings, and the first official "skyscraper" of ten stories or more was the 1885 Home Insurance Building in Chicago. From then on American office buildings simply grew taller and taller. The Chrysler Building took over the mantle of the world's tallest building from the Eiffel Tower in 1930 and the skylines of American cities became symbols of the wealth and engineering prowess of the country.

Skyscrapers also exploited another novel material: reinforced concrete. Modern concrete had been perfected in the middle of the nineteenth century by the father and son Joseph and William Aspdin, producing what they called portland cement. The process heats a mixture of limestone and clay to high temperatures—above 2,500°F—causing the ingredients to form a clinkerlike material that is then ground into a powder: cement. When water is added to the cement, the two react and set hard. If the cement is combined with sand or gravel, it produces a new substance—concrete—that sets with the properties of stone, but which can be poured and molded readily into almost any shape, rather than having to be laboriously hand carved.

Concrete is weak in tension, just like stone or brick, so on its own it could never replace timber. However, engineers found that when it was combined with steel, it could be made into a structure that withstood both compression and tension forces. The first such material that could do this was reinforced concrete, which is made simply by pouring concrete over a framework of steel rods. The steel rods reinforce the concrete in tension, while the concrete stops the steel rods from buckling when they are squeezed and also protects them from rust. The disadvantage of reinforced concrete is that the material only harnesses a small part of steel's tensile strength; if the material is put into tension, the concrete will crack long before the breaking point of the steel is reached, and this can allow water to enter the structure and corrode the steel. This problem led the German engineer C. E. Doehring to patent a new material, prestressed concrete. To build a prestressed structure, holes are drilled through the concrete, and thin wires are pulled through them and stretched taut after the concrete has set. This process puts the concrete in precompression, just as the wooden scaffolding hanging from the steeple at Salisbury Cathedral compresses the masonry below. The result is a material that has better properties than either concrete or steel on their own, one that withstands both tension and compression forces, and which can be made into beams that resist bending even better than wooden ones. Today the great majority of large public buildings and the pillars and lanes of most bridges are made from prestressed concrete, which allows them to be built quickly and cheaply.

By the middle of the twentieth century, the physical appearance of our world had changed beyond recognition from the start of the nineteenth century. Replacing timber with a new set of industrial materials had allowed engineers and architects to build ever-larger structures. We are nowadays dwarfed by our surroundings, especially in cities. I can remember the shock I felt on my first visit to North America, to Vancouver, Canada, glancing out of my hotel window and realizing I was looking down on the city's cathedral.

If the nineteenth century saw timber replaced in large-scale structures, the twentieth century saw the invention of new materials that replaced wood for many small-scale applications: plastics. Wood is perfectly good at making simple objects; pencils, toothpicks, and matches are all still wooden. The problem with making a more complex shape, though, is the time it takes to carve them, something that is simply not economic in our mechanized world. Some firms still make wooden toys, but they tend to be crudely carved and expensive, attractive only to nostalgic parents and grandparents. Children invariably prefer cheaper, more brightly colored and detailed plastic toys, and adults themselves prefer to buy cheap disposable toothbrushes, razors, and pens rather than expensive wooden ones. But it is no easy matter to replace such a light, strong material as wood. Late-nineteenth-century toymakers produced tin soldiers that could quickly and cheaply be cast in molds, but not until the 1920s was a new, easily moldable material invented that was as light as wood.

As oil became available in the twentieth century, chemists started to work to convert some of its heavier molecules into useful products; the Belgian chemist Leo Baekeland found that when two chemicals, phenol and formaldehyde, were heated together, they formed a resin that would solidify in a mold and produce items of any desired shape. The only problem was that though this material was stiff, it was alarmingly brittle. Baekeland found that he could overcome this problem by mixing the resin with wood flour to make a fiber-reinforced material; unwittingly he had copied the strengthening arrangement used in the cell walls of wood itself. The process was a success, and Baekeland found that he could produce plastic components of all shapes and sizes. The objects that were made in this new material, Bakelite, included a wide variety of new technologically advanced products: radios, telephones, and the like, whose rounded, fluid design had a great influence on the streamlined aesthetics of the art deco period. These items were not strong but were cheap to make and became wildly popular. Another type of fiber-reinforced material developed between the wars was cellulose-reinforced laminates, such as Formica. In these structures, the resin was reinforced by cellulose sheets in the form of paper, which could be decorated with attractive patterns. The sheets of material were then glued to wood or chipboard to produce shiny, easy-to-clean surfaces. The resulting product transformed people's lives, especially women's. Kitchen designs became brighter, and housewives no longer needed to scrub down wooden working surfaces, so hygiene improved dramatically.

The chemists never totally overcame the problem of the brittleness of thermosetting plastics such as Bakelite, so after World War II, materials scientists developed a range of plastics that did not need to be reinforced with fibers. In thermoplastics such as polyethylene and PVC, the long-chain molecules fold up as they set, producing much tougher materials. If you stretch a piece of polyethylene, for instance, the molecules unfold, absorbing large amounts of energy and helping stop any cracks running through it, one reason why it is often so hard to get into plastic packaging. The ease with which these thermoplastics can be molded into all manner of complex shapes has enabled them to take over for making most of the small objects that used to be made of wood or Bakelite. This has made our lives much easier, but it has had environmental consequences that we are only just beginning to appreciate: our rivers and seas are becoming full of nonbiodegradable plastic waste.

And a final type of fiber-reinforced plastic has been developed in the last eighty years that is even stiffer and stronger than wood, and which unlike iron and steel is just as light. These plastics contain continuous fibers that are even stiffer than the cellulose molecules found in wood. In fiberglass, the plastic resins are reinforced by newly pulled fibers of glass, and this material has been widely used to make the hulls of small boats and more recently the blades of wind turbines. However, the darlings of composite-materials technologists must be carbon fiber composites. In recent years these have started to replace not only wood but even steel because they can be made stiffer than both of them. Carbon fiber composites can be ten times as stiff as wood and are only half the density, so they are particularly useful to make structures where high performance is the highest priority. They are most commonly used in sports equipment; carbon fiber–reinforced rowing shells have replaced boats made out of plywood; carbon fiber skis have replaced the original wooden planks; and carbon fiber tennis rackets have replaced the old handmade wooden ones. Carbon fiber composites are even replacing steel and aluminum in the bodies of airplanes and racing cars. But there is still one problem with them: composites are not as tough as wood or steel, something that in the 1970s almost killed off the British engine maker Rolls-Royce. To keep down the weight of their new RB211 jet engines, the company developed carbon fiber turbine blades. They worked brilliantly, but would unfortunately shatter if a bird was dragged into the air intake, leading to catastrophic engine failure. Rolls-Royce had to return to using titanium blades, and the huge development costs for carbon fiber blades the company had incurred almost bankrupted it; Rolls-Royce had to be bailed out by the British government.

But despite a few setbacks, the new industrial materials have engineered a world that would have been unrecognizable to people of the eighteenth century. We dominate our surroundings as never before, but we also pollute our environment as never before. Surrounded by the giant structures outside our houses and myriad small objects inside, we are apt to feel overwhelmed by the world we have created. It is tempting to think that in this crowded material world there is no place for such an old-fashioned material as wood. Metals have superior stiffness, strength, and toughness and so have replaced timber in mechanical and civil engineering. Prestressed concrete is stronger and easier to form and so has replaced timber in architecture. Plastics are cheaper and easier to mold into the wide range of objects we need for everyday life. And fiber-reinforced composites have unrivaled stiffness, enabling them to take over the world of sport. Yet, in the twentieth century we were also able to exploit these new materials to make timber more useful than ever.

Wood was a key to the success of an industry that was only born at the start of the twentieth century: building airplanes. You might think that such advanced machines as airplanes should have been constructed from the very start using "modern" materials such as metals, which after all are stiffer and stronger. However, for the early airplanes, which had low-powered engines, the key requirement for the airframe was for it to be as light as possible. As we have seen, weight for weight, wood is just as stiff as metal, and because the same weight of wood can be made into a thicker strut, it will actually be stronger and more rigid than an equivalent metal one. The early aircraft were therefore constructed using a rectangular box-work of wooden struts, joined by metal plates, and with the frame braced by diagonal wires to resist the shear forces. In some of the earliest planes, the Wright Flyer and Blériot monoplane, for instance, the framework was left open to full view, except for the wings, which were covered in fabric. Later, though, the fuselage was also fabric covered to make the machine more streamlined. World War I biplanes such as the Sopwith Camel and its adversary the Fokker Triplane were miracles of lightweight design, yet were churned out in the thousands, often by sub-contracting furniture makers.

Early-airplane manufacturers were also major users of the first truly novel material to be made out of wood—plywood. As we have seen, the anisotropy of wood—the huge difference between its properties along and across the grain—has always made it tricky to shape and to use. Wood tends to split along the grain if loaded in the wrong direction, often with catastrophic results. The idea of overcoming this disadvantage by gluing together several thin sheets of veneer so that the grain of adjacent sheets is at right angles to each other was first suggested in 1797 by the foresighted Samuel Bentham. As he pointed out, such sheets should prove particularly useful in building boats, as they would withstand shear forces far better than the traditional plank-built hull. In his time, though, plywood proved impractical because veneers could only be cut by sawing thin sections of wood from tree trunks. Plywood sheets made using this method would have been as limited in width as conventional planks. This problem was only overcome in 1851, when Immanuel Nobel, father of Alfred Nobel, the inventor of dynamite and benefactor of the eponymously, if ironically, named Peace Prize, patented an effective rotary lathe. A log is mounted at both ends in this huge machine tool and spun around while a long knife is applied all along its length, cutting through and unraveling a thin sheet of veneer, like a sheet of toilet paper being unwound from a roll. The technique requires the log to be almost perfectly straight, but for the first time this meant that huge sheets of veneer could be produced. At first, the quality of plywood made from these veneers was poor because the animal and plant glues that were used to stick the veneers together were not waterproof and tended to rot. Plywood was largely limited to internal use and for simple items such as artists' boards for oil painting. But furniture designers soon found it to be an ideal material. Designers such as John Henry Belter of New York showed that it could readily be bent and molded into two-dimensional, and even three-dimensional, curves, enabling him to produce graceful rococo revival furniture.

The ability to make curved shells also attracted early airplane designers as it suggested a way in which they could build smoother, more streamlined fuselages for their aircraft. In 1912 a monocoque Deperdussin monoplane, piloted by Jules Védrines, won the Gordon Bennett Trophy race. This graceful midwing monoplane, with its conical fuselage made from three-ply tulipwood sheets, and with a huge streamlined propeller spinner, was way ahead of its time, pointing the way for the airplanes of the future. In World War I, which followed soon after, though, the plywood design was only taken up by a few German designers. Small numbers of beautiful Roland C.II "Walfisch" reconnaissance aircraft, my personal favorite, and the Albatros D.V fighter were built, and they looked great, the front of the fuselage often being painted with the jaws of sharks, to emphasize their sleekness and ferocity. However, they were outnumbered by planes with the traditional braced wood frame. The problem may have been the slightly higher weight of these aircraft or their tendency to rot in wet weather.

In the 1920s the waterproofing problem was finally overcome by gluing the veneers together with new polymer-based resins. Plywood for indoor use was glued using cheap urea-formaldehyde resins, while plywood to be used outdoors was made with waterproof phenol-formaldehyde glues. Soon eight-foot-by-four-foot sheets of plywood were being mass-produced for general building use, and it became the material of choice for professional carpenters and unskilled do-it-yourself enthusiasts alike. Manufacturers also developed high-quality marine plywood, which has been used to make all sorts of small boats. In particular in the last sixty years, plywood construction has made dinghy sailing much more accessible, through the design of cheap kit-built dinghies. Over seventy thousand of the two-handed British Mirror dinghies have been built and four thousand of the single-handed South African Dabchick. My father spent many weekends in our garage building a plywood Minisail in the late 1960s, though I seem to recall that the construction phase was not all plain sailing. Marine plywood's ability to withstand moisture has even been exploited by musical-instrument makers. I remember a particularly enthusiastic owner of a plywood bass recorder whom I met at an early music summer school. Being square in cross section it looked extremely ugly, but it certainly sounded as good as any bass recorder I have heard, which is not very many.

And despite the increasing power of air engines and the resulting increase in the speed of airplanes, which promoted the use of stiffer, if heavier, metals such as steel and aluminum, plywood made a dramatic comeback in World War II. Worried by shortages of metals, British aircraft designers developed cheap wooden gliders for airborne assaults such as the Normandy landings, and even a superfast bomber. Nicknamed the Wooden Wonder by its crews, the de Havilland Mosquito had a fuselage made using an ingenious sandwich construction, in which a light central core of balsa was covered on each side by three-ply birch plywood. The shape was formed by molding the two halves separately and then gluing them together. The wings were also covered by sheets of plywood. Powered by two Rolls-Royce Merlin engines, the Mosquito was faster than most German fighters, and almost eight thousand were made. They operated in a wide range of roles from reconnaissance aircraft to light bombers and even to night fighters. The construction technique was so successful that de Havilland even continued making the nacelles of their postwar fighter jet, the Vampire, from plywood.

Plywood is still an important material, and over 190 million cubic yards of it is made every year, but in the last seventy years it has been joined by a wide range of other "engineered wood" materials. Most of these have inferior properties and are largely methods of using up otherwise worthless leftover cuts of wood. The simplest is chipboard, which is made by gluing wood chips and sawdust together with resin glues, and pressing them into sheets. Though weak, chipboard is often used as the central sandwich layer in cheap furniture, such as IKEA's iconic BILLY range of bookcases, where it is covered with veneer, which strengthens it and improves its looks. Forest-product laboratories in the USA and Scandinavia have also developed a wide range of fiberboards, made from wood that has been mechanically macerated to separate the individual wood cells. These are then pressed into sheets and waterproofed with resin glues. Fiberboards have a wide range of uses, from high-density hardboard, through medium-density MDF, the darling of the interior decorators you see on TV makeover shows, through to the lightest, cardboard, which has proved its worth as a cheap packaging material. Today around 330 million cubic yards of these materials are made every year.

But the fastest-growing use of wood today is in producing materials that have better properties than timber or even plywood: laminated wood or glulam, and crossed-fiber laminates or CFL. Wood laminates are made from sawn planks of wood, so they look beautiful, but the big advantage is that by gluing together large numbers of cheap, short planks, one can make wooden beams and sheets of almost any size and shape; architects are no longer constrained by the size of trees. The big breakthrough in making wood laminates was achieved by the invention of finger joints. The ends of each short plank are machine cut into wavy ends, like the fingers of a hand, which precisely fit into each other; when these are glued together, they form a joint almost as strong as the wood itself. To make a glulam beam the individual planks are simply laid out side to side and one on top of the other, in any number of layers, with the finger joints staggered to avoid weakening the beam. The beam can even be bent at this stage, like a pack of cards, before the whole structure is glued together and pressed with hydraulic machines. Crossed-fiber laminates are made in just the same way, except that the alternate layers of the planks are oriented at right angles to each other, as in plywood, to make the sheet equally strong in all directions.

Laminated wood has many uses, including making attractive furniture that combines the strength and low cost of plywood with the beauty of timber. But the most spectacular use of this material is in architecture; designers can create beams of all shapes and sizes, whose shape can be computer controlled. The beams can then be joined together to form the structure of a building using steel bolts and plates. The result has been a flowering of building in wood, adding new lightness and grace to the homely feel of traditional timber. In Britain, the material has often been used in the design of beautiful garden buildings.

The Winter Gardens in Sheffield, for instance, is a new take on palm houses, its glazed external shell being held up by graceful parabolic arches made from laminated larch. The visitor center at the RHS Garden Hyde Hall in Essex is a modern reinterpretation of the old tithe barn, but with huge laminated pillars and rafters, and a glazed nave to let in more light. But the full artistic and engineering possibilities of these materials have been explored elsewhere in the world. The Centre Pompidou-Metz, in Lorraine, France, an outpost of the Centre Pompidou in Paris, is perhaps the most spectacular example of the use of glulam's artistic possibilities. The roof is a sweeping three-hundred-foot-wide hexagon, supported by a hexagonal framework made from ten miles of laminated beams, and rising to a central spire 253 feet high. In the USA and Canada the engineering advantages of the lightweight glulam beams have been exploited in the design of huge sports halls. Eastern Kentucky University's Alumni Coliseum has the world's largest wooden arches, spanning over 308 feet, while in the Richmond Olympic Oval, built for the 2010 Winter Olympic skating events, the roof is held up by 3,100 cubic yards of Douglas fir glulam beams. Both of these structures are comfortably wider than the wrought-iron St Pancras arch.

Perhaps the most exciting development of wood-laminate architecture is in the construction of a new generation of high-rise buildings. In the last few years, wooden buildings have been getting taller and taller, from the Forte tower in Melbourne, Australia, which reached a record height of 106 feet and ten stories in 2012, to the current record holder, the eighteen-story, 280-foot Mjøstårnet mixed-use tower in Brumunddal, Norway. Held up by thick wooden beams and coated with thick plates of CFL, these huge skyscrapers weigh only around a fifth as much as conventional concrete-and-steel structures, have only around a half of the embodied energy, and despite wood's reputation for flammability, are better at resisting fires; the huge beams simply char at their surface, cutting off access to the interior to the flames, whereas steel frames are all too apt to melt and collapse. Even taller wooden buildings are being planned, from a twenty-one-story block in Amsterdam, a forty-story tower in Stockholm, and even an eighty-story tower in the Barbican, London. Look around and you may soon see a tall wooden tower looming above you.

And wood also acts as the feedstock for a range of industries all based on one of the original wood-pulping processes. In 1930 G. H. Tomlinson developed a recovery boiler that recycled all the inorganic pulping chemicals used in the kraft process, making it far more efficient. Nowadays almost all pulping is done using the kraft process, and over 440 million tons of paper are annually made. Of course, not all of it is for the book trade or for newspapers. In the last eighty years, technologists have developed a wide range of paper products, from packaging through toilet paper to sanitary products. A good deal of wood pulp is also further processed to produce pure cellulose, which acts as the raw material for a range of fibers, sheets, films, and lacquers. Cellulose was first used to produce the nonflammable dopes that coated the fabric of the early wood-framed aircraft we looked at earlier in this chapter, but a whole industry developed in the 1920s and 1930s that was based on wood pulp. The fiber viscose is produced from cellulose xanthate, while cellulose acetate and nitrate are used to produce a range of plastics, and cellulose triacetate is the basis of cinema film. And paper itself, which the computer promised to consign to the dustbin of history, is used in ever greater quantities. The paperless office remains a pipedream, and no workplace is without its banks of printers and photocopiers. Even e-readers have failed to take over from hardback and paperback books as had been predicted. People seem to love the convenience and tactile qualities of the printed book.

Plainly, wood can no longer be regarded as simply an old-fashioned and obsolete material. We have used the same industrial methods that we developed to produce other competing materials to fashion wood into a whole new range of products that compete well with modern metals, concrete, plastics, and composites. Wood production and use is growing year by year, and from around 1.9 billion cubic yards in 2018 it is likely to rise to around 2.2 billion cubic yards by 2030. A greater volume of wood is now used even than that of its nearest rival, cement, whose production is currently estimated at 1.7 billion cubic yards per year. But our demand for wood and our use of trees continues to exert huge pressures on the growth of forests and trees around the world. We shall see in the next chapter the extent to which our relationship with wood has affected the environment and altered our planet.

PART 4

FACING THE CONSEQUENCES

CHAPTER 14

Assessing Our Impact

onsidering how much wood we've always used, it would be surprising if our relationship with trees hadn't affected our history in some way. But though historians have started to consider several related subjects—the role of energy and coal in the industrial revolution, for instance—for the most part they have ignored the role of wood. It's certainly not a subject that lends itself to the usual historical method of poring through archives. The foresters and carpenters who dealt with wood and who were the real experts were generally humble folk who have left few written records, while the few people who did write about wood, such as Pliny the Elder and John Evelyn, were aristocrats and gentlemen with no real hands-on experience. These writers were apt to be swayed by their personal experience of seeing trees being felled, a highly dramatic process that is far more memorable than the almost imperceptible growth of the trees around us that occurs throughout our whole lifetime. It is not surprising, therefore, that the tales that popular histories tell are generally ones of destruction; the writings are full of complaints about how forests have been "razed to the ground" and woods "spoiled." Consequently, books are full of deforestation myths, morality tales that relate how, as a new power rose to prominence, it overexploited its forests for timber, and how the resulting deforestation led to environmental catastrophe—soil erosion, climate change, famine, and eventual collapse. These deforestation myths have cropped up again and again throughout history; deforestation has been blamed for the fall of several ancient Mesopotamian empires, Mycenaean Greece, the Mayan Empire, the Venetian Republic, and, most devastatingly, the collapse of civilization itself among the Rapa Nui of Easter Island. Perhaps the most frequently cited example of all is the foundation myth of the British Empire: that the construction of the Royal Navy destroyed the country's great primeval oak forests.

The truth is very different. We have certainly had a huge effect on the world's forests: reducing forest cover and changing the composition of the forests that do remain. However, people have found ways of coping with the loss of forest, maintaining an adequate supply of wood while avoiding environmental collapse. But as we shall see, our impact has been pervasive at all scales right up to the global one, so that our relationship with trees has had a profound effect on world history.

The deforestation myths, though superficially attractive, are based on false assumptions. The first is that felling trees results in catastrophic soil erosion. Once again people have been swayed by the immediate evidence of their senses, and an experience of modern industrial forestry practices. After forests are cut down, the streams and rivers that they drain into quickly cloud over with muddy water. But our impressions exaggerate the historical extent of the problem. In the past, woodland would have been cleared far more slowly and without heavy modern machinery would have damaged the soil far less. The immediate effect of deforestation would have been far smaller.

I have had some experience of the reason why so many people exaggerate the pace of erosion, especially in tropical rain forests, where the soil is particularly delicate, the heavy logging machinery is particularly destructive, and the rainfall particularly heavy. These factors might be expected to cause rapid soil loss. When I was at the Danum Valley Research Centre, Sabah, Borneo, in the early 1990s investigating why tropical rain forest trees develop their huge buttress roots, two other research groups were investigating the effects of deforestation on soil erosion. One was looking at erosion at a small scale, on research plots around a logging camp; the other was looking at erosion on a larger scale, measuring the amount of soil swept from the whole catchment area draining into the river Segama. The survey on small cleared plots sixty-six feet long by seven feet wide showed that a single massive storm could remove 200 pounds of soil, a depth of about 0.06 inches, which would have severely damaged the soil. In contrast, though the same storm washed sixteen thousand tons of silt into the Segama, this was over a catchment area of 1,150 square miles; a simple calculation showed that this represented an average depth of soil of only 0.00016 inches, four hundred times less than the soil loss from the small plots. The cause of this mismatch was that the soil lost from the small research plots had merely been shifted a few yards downhill; though the plots had lost soil, they would also have gained it from areas farther up the slope; little soil had actually been lost from the forest and made it into the river.

Once land has been cleared for agriculture, it *can* speed up erosion. Archaeologists often find that the topsoil at the bottom of slopes is somewhat deeper than at the top as the soil has slowly crept downward. However, these changes have taken place over hundreds or even thousands of years; erosion actually occurs only slowly. Even in the one type of soil that is known to be particularly vulnerable to erosion, silty loess soils, the process is hardly an overnight phenomenon. In central China, the colored loess soils that give the Yellow River its name can erode at a rate of up to 0.04 inches a year, and the windblown soils cause the notorious dust storms that can choke huge areas of northern China. However, though high pressure on this land has progressively led to some desertification, this only started to occur after the area had been intensively farmed for thousands of years.

People have also learned to manage soils for millennia so as not to unduly damage them. The roots of grasses and crop plants can be just as good at holding on to soil as tree roots. In temperate regions, with their mild climates and moderate rainfall, the fertility of soils can also readily be maintained by judicious manuring and composting. Only in upland areas where rainfall greatly exceeds the evapotranspiration of water from crop plants can soils become waterlogged, acidic, and unproductive. Similarly, though nutrients are more likely to be leached by the heavy rain in tropical rain forests, recent research has shown that the inhabitants of the Amazon and West African rain forests devised a whole host of ways of maintaining yields in their farm gardens. They constructed raised terraces and fertilized them with large-scale use of manuring, to produce what are known as Amazonian Dark Earths.

Around the world, farmers also tend to avoid the steeper slopes where erosion would occur most rapidly and where storms can cause catastrophic mudslides. Even today, steep river gorges and scarp slopes tend to be covered in remnant semi-natural woodlands. In the most highly populated and hilly areas, where demand for land meant that people had to farm even the steepest slopes, farmers traditionally resorted to terracing to prevent soil from washing down the hill. This technique was particularly common in the limestone hills of the Mediterranean, the volcanic Canary Islands, and in the rice paddies of Southeast Asia. Whole landscapes in these regions were carved into terraces that still hug the sides of the hills, like the contour lines on a huge map, holding the soil behind their retaining walls.

In one respect only has soil erosion had a major effect: on the rivers that drain deforested areas. The displaced soil not only discolors them, but can also silt them up, causing trouble downstream. The fens of East Anglia were created in Neolithic times, largely because soil from deforestation upstream blocked up the rivers that drained into the Wash. And throughout history, people have had trouble keeping their harbors open; silt washing downstream tends to be dropped at the mouths of estuaries, blocking the channel. Most estuaries have to be continually dredged to keep their shipping lanes open and in extreme cases the silt can build up into land, isolating the harbor from the sea. For example, the city of Ephesus, once a major port of the ancient world, had its harbor blocked by silt from the Cayster River and was stranded several miles inland, finally being abandoned in the fifteenth century by the Ottoman Turks.

A final piece of evidence that soils are not usually damaged by deforestation is the strong tendency of land to revert to forest; clearing an area of woodland does not mean that it is lost forever. As any farmer, nature-reserve manager, or gardener knows, it is a constant battle to prevent trees from invading and forest reestablishing itself. Any land left fallow quickly gets colonized by tree seedlings and reverts surprisingly rapidly, first to scrub, and finally to forest, a process that ecologists call secondary succession. The results of this process can perhaps be most clearly seen by the wealthy commuters of the outer suburbs of New York and London. In the last two centuries farmers have abandoned their small farms on the unproductive soils of Connecticut and Surrey, letting them revert to the secondary woodland that now covers the land around the commuters' dormitory villages.

And the deforestation myths themselves do not stand up to closer examination. Take one of the most cited myths, the environmental destruction of Easter Island. The Rapa Nui did indeed clear the island of its forest cover, but this was a perfectly rational thing to do. The palms that dominated the forest could not have produced useful timber, since palms are monocotyledonous plants that do not make proper wood. And the shed fronds of palms smother the forest floor, forming an infertile blanket that prevents the growth of any understory plants. By removing the palms, the Rapa Nui successfully converted their forests into fertile farmland, sheltering their crop plants from the wind using stones, and creating gardens in volcanic hollows. They were thriving when the first Europeans arrived in 1722, only to be decimated by European diseases. The real environmental damage only occurred in the twentieth century following the importation of sheep by Chilean speculators. They swiftly eradicated most of the remaining native vegetation, causing damaging soil erosion. It is the uncontrolled grazing of sheep that has turned Easter Island into the same sort of green desert that they have long since made of Britain's national parks and of mountains around the temperate world.

● ● ●

The simple narrative that deforestation inevitably leads to disastrous soil erosion and environmental collapse is therefore simply not tenable. But our relationship with trees *has* had a huge effect on civilization and on the global environment. We can obtain a far better understanding of the historical geography of civilization and understand our present situation much better if we take a more realistic biological perspective on our relationship. In particular we need to take into account that not all trees are the same; there are large differences in the biology of the two major taxonomic groups of trees: broad-leaved trees, and conifers.

Broad-leaved trees have more efficient water-conducting vessels than conifers and can grow faster as seedlings, so they outcompete conifers and dominate the vegetation in areas with the best climates and soils. They tend to have large, branching crowns and curved branches, and many species resprout from their stump or regrow from their roots if they are cut down. Broad-leaved forests also tend to improve the soil; the leaf litter that they drop builds up its fertility. In contrast, conifers are naturally restricted to areas with poor growing conditions, places where frost and drought kill broad-leaved trees—the cold northern reaches of the globe, high mountains, and semidesert— and areas with poor, thin soils; and their needles acidify the soil and lock up what little nutrients are present. They also have a much more limited form of branching than broad-leaved trees, so their trunks are much straighter, knot-free, and uniform. And most species die when they are cut down; any regeneration has to come from seedlings.

In recent years, environmental historians have combined this biological perspective with techniques that collect empirical evidence about the distributions of trees and are starting to discover what really did happen and why. Looking at old maps and parish records allows them to identify changes in tree cover over the last thousand years, and pollen analysis can help quantify changes in the extent of tree cover and the species composition of forests even before written records were available, right back to the end of the last ice age. When we use their evidence and place the biology and ecology of trees center stage, world history suddenly makes much more sense. We can see how forests affected us, and how in turn we affected the ecology and environment not only of the world's forests, but of the whole planet.

Over time the same patterns kept recurring all across the world. The most obvious one is that whenever farmers colonized new regions, the first places that they settled in were those that had previously been covered in broad-leaved forest; it was simply the best, most productive land. They did not cut the trees down for firewood or timber, but to clear the land so that they had space to plant their own crops. In Europe the first areas that Neolithic farmers cleared were the evergreen forests of the Mediterranean, and the deciduous forests of central and Western Europe. In Asia, they cleared the deciduous forests of central China, and the evergreen forests of southern China, India, and Southeast Asia. In the Americas, farmers cleared the monsoon forests of southern Mexico and the Caribbean, and the inhabitants of Amazonia even started to clear and farm land in this huge area of rain forest. In Africa, farmers cleared the rain forest of West Africa, and the savannas of central, eastern, and eventually southern Africa. In more recent historical times, the Spanish settlers in the Canary Islands cleared the evergreen laurel forests from the hills, while the first land in North America that the Europeans cleared were the deciduous woodlands of New England and the Mediterranean evergreen forests of California's Central Valley. In New Zealand, the Maoris, and later the British settlers, preferentially cleared the lowland southern beech forests.

In contrast, for the most part farmers avoided conifer forests because they were an excellent indication that the land was too poor to farm. Neolithic farmers, therefore, left the conifer forests of Scandinavia, the Alps, Siberia, and northern Japan largely intact, while in the New World the conifer forests of Canada, the Great Lakes, the Deep South, and the Pacific Northwest remained largely untouched until the nineteenth century. Indeed early European settlers to North America were given specific advice when they were settling a new area to clear only the broad-leaved forest and leave the conifers standing. In Ohio, for instance, beech was a clear indication of good soil conditions.

The consequence of this pattern of settlement was that the wealthiest, most stable, and longest-lived states grew up in areas formerly dominated by broad-leaved forest. The great civilizations of Ancient Greece, Rome, and China were all based in such areas. It was the states on former grassland and desert regions, states that were fed by irrigated land, such as the Mesopotamian civilizations of the Middle East, the Anasazi sites of New Mexico, and the many civilizations around the Andes, that tended to collapse. Not only were they more vulnerable to droughts, but the irrigation water brought in salts that were left as the water evaporated and built up in the soil, causing salination and dramatic long-term falls in crop yields. For this reason agriculture on the Great Plains of America is unlikely to be sustainable in the long term, and the area may well return to the dust bowl conditions of the 1930s. The irrigated land of Ancient Egypt was an exception because the annual flooding of the Nile not only irrigated the land but also washed away the salts; this is the reason why the great Egyptian civilization lasted for thousands of years.

One potential problem faced by farmers living in former broad-leaved forest was that in clearing the forest they destroyed their supply of wood and timber. This did not prove overwhelming, though. Farmers simply kept a small area of forest and managed it for firewood and timber, exploiting the ability of broad-leaved trees to respond to damage by resprouting. In Southern Europe, trees were often shredded; people cut off their lateral branches every few years, leaving just the trunk and upper canopy, which would then rapidly resprout and produce a new harvest of wood. In Northern Europe and Japan, farmers managed broad-leaved forests by coppicing them. They cut the trunks of trees down to just above the ground every ten to twenty years, stimulating them to produce large numbers of fast-growing shoots. And right across the globe people also kept domestic animals such as cattle, horses, and pigs in their woodlands, cutting the trees above the browse line to prevent the animals from damaging the regrowing shoots, a process known as pollarding. The animals could still graze the grass of the forest floor and guzzle the autumn bounty of acorns and beech masts, and shepherds could feed their beasts on branches cut from the trees. Using these techniques, farmers could supply all their needs for firewood while setting aside only a small proportion of their land as coppice or wood pasture. And just by allowing a few of their trees to grow naturally to maturity, they would have had more than enough timber. Paradoxically, the need for wood meant that broad-leaved forest survived even in densely populated and intensively farmed areas.

As these civilizations flourished, their demands for firewood, charcoal, and timber all rose. We saw in chapter 10 that limits to the supply of firewood did eventually constrain their economic growth, since it was just too bulky to transport it cost-effectively to distant cities. However, the same was not true for timber, and from the Bronze Age onward a timber trade developed. Large timbers were moved both within and between states. Areas and states that had extensive conifer forests could supply the wealthy deforested regions with the large structural beams they needed for their major buildings and the timbers they required to construct their ships. The Phoenicians of Tyre and Sidon exported their famous cedars of Lebanon to Egypt, for instance, and most famously they sold Solomon the beams he needed to roof his temple. In the northern Mediterranean, Macedonia, Lebanon, and Thessaly exported firs to ancient Athens to build up its fleet of triremes. Softwood logs were rafted down from the Dolomites to supply the shipbuilders of the Venetian Republic. And in Northern Europe, the Baltic nations exported spruce, fir, and pine to the cities of the Hanseatic League and onward to Holland and England. The trade was a way in which poor, sparsely populated areas could develop, obtaining the foreign currency and food they needed from more densely populated and wealthier ones; it was a way in which wealth was spread across Europe. It's notable that several of these softwood-exporting countries later became powers in their own right: the Phoenicians developed an empire of trading ports around the Mediterranean, including the great city of Carthage; the Macedonians, under Philip and his son Alexander the Great, conquered the rest of Greece and eventually most of the known world; and Sweden rose to prominence in the seventeenth and early eighteenth centuries, briefly dominating a Northern European empire and challenging the great powers of central Europe.

In the tropics, the hot, humid climate and high rainfall speeds up tree growth, hinders travel, and makes hard labor more difficult. Hence, even in seasonally dry monsoon forests, it has always been harder to clear forests and manage the land. Consequently, when the Europeans started to colonize tropical and subtropical areas from the sixteenth century onward, they treated them very differently from temperate regions. Europeans had rapidly settled and farmed those areas with a similar climate to that of Europe, building colonies that were essentially copies of countries in the Old World: Canada, Australia, New Zealand, South Africa, Chile, and Argentina. In contrast Europeans showed no desire to farm themselves on the former monsoon forests they were logging for expensive timbers such as mahogany and teak. Instead, the ruling elite chose to establish large plantations on which they could grow cash crops. In the Caribbean, sugar and cocoa were the preferred crops; in Brazil, coffee; in the Southern American colonies, cotton and tobacco; in India, tea; and in Southeast Asia, rubber. And rather than working the land themselves, the colonists employed a workforce that could better withstand tropical diseases. In India and Africa these were local people, but having exterminated the indigenous populations of the New World, colonists simply replaced them with a new workforce. They shipped in slaves from Africa or indentured laborers from the Indian subcontinent and forced them to work in the fields of Brazil, the Caribbean, and America's Deep South. Thousands of Indian and Chinese laborers were also imported into Southeast Asia. This resulted in unprecedented mass movements of people across the globe and rapid deforestation of much of the lowland tropics. The island of Barbados, for example, though named after the bearded fig tree, *Ficus citrifolia*, a notable component of the forest that once cloaked it, today has a forest cover of less than 5 percent. Run by a small number of colonists, the huge plantations of the tropics and subtropics sent back vast quantities of raw materials and luxury food to the industrialists of Europe and in turn acted as a market for the industrialists' finished goods.

In the last century and a half, industrialization and population rise have increased our demands for wood and for land even further; and the rise of technology has allowed us to satisfy it by cutting down and clearing unprecedented areas of forest. The first areas to fall to industrialized logging operations were the great conifer forests of North America. In the late nineteenth and early twentieth centuries, first the white pine forests of the Great Lakes area, then the slash pine forests of the Deep South, and finally the Sitka spruce forests of the Pacific Northwest were clear-felled by huge forestry corporations, most notably the Weyerhaeuser Company, which shipped its timber to the industrializing Northeast of the country via the newly opened Panama Canal.

Since 1940, the center of destruction of forest for timber has moved to the tropics, where for the first time powerful chain saws and modern haulage equipment has allowed loggers to penetrate and log the previously inaccessible rain forests of central Africa, Southeast Asia, and Amazonia. There the selective felling of valuable timber trees has resulted in widespread damage to the forests. This in itself would not be disastrous. Rain forests recover from damage even faster than temperate forests, and logged areas are swiftly recolonized, first by huge-leaved pioneer trees such as balsa, *Cecropia* and *Macaranga*, and eventually by the climax species. Indeed, it is only the discovery of telltale crop plants in rain forests and the recent finds of Amazonian Dark Earths that told us that many areas of apparently virgin rain forest had been cleared and farmed in previous centuries. The main problem of recent logging operations is that logging roads have opened up the rain forests, facilitating the same sort of clearance for agriculture that in Europe happened four thousand years ago, and that in the Caribbean and North America took place in the eighteenth and nineteenth centuries. The forests are being cleared at an unprecedented rate to make way for subsistence farming and on a much larger scale for cash crops such as oil palms and soybean and for cattle ranching to provide cheap beef.

Now that we can understand the pattern of where, how, and why we have exploited trees and forests, we can make better sense of the effects of our actions, both regionally, on the forests themselves, and globally, on the ecology of the whole planet. Environmental historians are starting to realize that even before industrial times our impact was immense. The most obvious effect was the reduction in the area of woodland, particularly in regions where broad-leaved trees had been dominant. In mid-latitude Europe, for instance, pollen analysis shows that tree cover fell from a maximum of around 80 percent six thousand years ago to 60 percent three thousand years ago and 40 percent by the end of the Middle Ages. In the mild climate of England, forest loss was even greater and occurred even earlier. Oliver Rackham showed from historical records that the forest cover of England, a country formerly dominated by broad-leaved forests, had fallen to as low as 10 percent at the time of the *Domesday Book*, in 1086, and fell even further, to around 7 percent by the beginning of the fourteenth century. In China, forest cover had fallen to around 20 percent by the fifteenth century. In sharp contrast, pollen records show that the loss of conifer forests in Scandinavia and the Alps was undetectable before two thousand years ago and barely noticeable even three hundred years ago.

A second effect was that the area of pristine "wildwood" or "old growth forest" declined drastically. Europe currently has just a handful of untouched sites, and the few that remain, such as the famous Bialowieza Forest on the borders of Poland and Belarus, are continually under threat from logging. On its own, this loss of pristine forests would not necessarily result in any great loss of diversity or of the ecosystem services provided by the forests, and there does not appear to have been any extinction of tree species. However, managed woods contain none of the truly giant trees that once dominated Europe's forests; early European settlers in America were astonished at the size of trees in the new continent. The trees that regrow after logging are shorter and have narrower, more crooked trunks than their predecessors and so yield inferior timber. The complexity of forest ecosystems and the amount of carbon they store is also lower, something that is especially true of coppice woodlands. Thousands of years of woodland management have also drastically altered the species composition of the remaining woodlands. For instance, pollen analysis has shown that in southern England forests in pre-Neolithic times were dominated by lime trees; oak and hazel dominated the north, hazel and elm in Ireland, and birch and pine in the Scottish Highlands. As lime does not coppice, and its wood is pale and soft, the early farmers did not find it particularly useful; they encouraged the growth of other tree species instead, and lime quickly went into decline. Centuries of selection for trees that supplied useful timber and firewood resulted in the massive spread southward of oak and hazel. Ash, whose coppice poles are used for tool handles, also became more common, while the Romans introduced beech and chestnut to the southeast. The modern composition of the woods of Britain is therefore very different from that of the original wildwood; so the great "primeval" forests of oak that were supposed to have been cut down in the eighteenth century to build the Royal Navy had never existed.

By combining the loss of woodland area with the reduction in the size of the trees and plant biomass, environmental historians have also been able to calculate the effect that we have had on the amount of carbon stored in forests. The forest historian Michael Williams of the University of Oxford has estimated that by 1700, some 1.5 million square miles of broad-leaved woodland had been lost, almost 3 percent of the total area of dry land and 10 percent of the global tree cover. According to William Ruddiman from the University of Virginia, this would have released something in the region of 275 gigatons (Gtons) of carbon into the atmosphere, which would have increased CO_2 levels by around 40 parts per million (ppm). Ruddiman has argued that, along with increased emissions of methane from rice paddies and from domestic ungulates, this would have raised mean air temperatures by 1.4°F. This was enough to prolong our present interglacial and stop the world from turning to a new ice age. And it is worth noting that this temperature rise is similar to the measured rise in temperature over the last century, which was caused mostly by massive burning of fossil fuels. Ruddiman's figures have been hotly debated, and other palynologists give the release of carbon at more like 110 Gtons, but the likelihood that deforestation might have contributed to global warming well before the Industrial Age is backed up by more recent research that has investigated the effects of historical events on forest cover, CO_2 levels, and climate. Pollen analysis, for instance, has shown that the forest cover of Europe has increased substantially twice in the last two thousand years: once, in the first few centuries after the fall of the Roman Empire in AD 400; and again at the end of the fourteenth century after the death of a third of the population from the Black Death.

And another major historical event is now thought to have helped cause the Little Ice Age, a period of global cooling that lasted throughout the seventeenth and eighteenth centuries. Environmental historians are finding that the discovery of the New World in 1492 did not initially result in *deforestation* of the continents, as was once thought, but extensive *reforestation*. Infectious diseases such as measles, influenza, and syphilis that were brought in by the Europeans decimated the indigenous people. From a population of as high as 60 million people, it is thought that only around 6 million were left by 1600, a phenomenon known as the Great Dying. The result of this depopulation was that the cultivated fields of Mexico, the terraces of the Inca highlands, and the forest gardens of the Caribbean and Amazonia all reverted to forest, an effect that reduced global atmospheric carbon levels by around 9 Gtons. This would have lowered the CO_2 concentration by around 3.5 ppm, enough to explain two-thirds of the global cooling of 0.27°F that was seen for the next two hundred years. The poor harvests of Europe in the seventeenth and eighteenth centuries that led to so much civil strife, and eventually to the French Revolution, were caused by the regrowth of forests thousands of miles away.

• • •

If our effect on the world was huge even in preindustrial times, since we have industrialized, our effect on the world's forests has accelerated rapidly. Between 1700 and 1940 a further 1.5 million square miles of woodland was lost in temperate regions, much of it conifer forest. And since 1940 we have been removing forty-five thousand square miles of tropical rain forest a year, a total loss of over 3 million square miles. The current world forest cover has fallen to 31 percent of the dry land, from around 43 percent six thousand years ago. The resulting carbon dioxide emissions are thought to be contributing around 20 percent of the total increase that is currently causing global climate change.

The rapid loss of forests did promote concern and action among foresters. The stimulus of industrial logging led to the invention of a new industrial way of managing trees, plantation forestry, a technique that was first developed in Germany in the nineteenth century. The idea was that instead of allowing coppiced broad-leaved trees to resprout, or conifer forest to regenerate naturally from seedlings, the best way to grow trees more quickly and more profitably is to clear-fell forests and replant them with young trees grown in nurseries. Essentially it is tree farming. This new "scientific" method quickly supplanted traditional techniques such as coppicing, which woodmen had successfully used for centuries. It rapidly spread from Germany to the rest of Europe, and eventually, in the twentieth century, to North America and the tropics.

Unfortunately, in many ways the spread of plantation forestry has proved disastrous. The first problem is that it assumes that timber is the only worthwhile resource; therefore foresters promote the planting only of fast-growing and straight-trunked trees: conifers, *Eucalyptus*, and teak. Huge areas of land have been cleared of native broad-leaved trees and planted with these species, reducing biodiversity. In warm temperate areas, many broad-leaved woodlands have been destroyed and the land covered with conifers, which reduces the soil quality, while the perpetual darkness beneath their canopy kills off the understory of woodland shrubs and flowers. In Iberia, huge forests of *Eucalyptus* have spread, outcompeting native forest vegetation and promoting forest fires.

Another problem is that plantation forestry promotes the growth of huge monocultural stands of a single species. These are particularly vulnerable to wind damage and, more worryingly, fungal diseases and pests, which can destroy whole forests. Foresters can do little about this because trees have too long a life cycle for selective breeding to develop new disease-free strains. Trees are simply not suitable for conventional scientific farming. The situation is made worse by foresters' love of using exotic trees that are particularly susceptible to diseases and damaging to local flora and fauna. For instance, the Monterey pine, *Pinus radiata*, a rare species from the coast of California, has been widely grown across the world, from South Africa to Chile, Australia, and New Zealand. I have even seen it grown in the Canary Islands, which have a perfectly good species of their own, the canary pine, *P. canariensis*, which is beautifully adapted to the local conditions. Moreover the exotic species introduce new pests and diseases that can kill native trees, which are not adapted to deal with them. This represents probably the biggest threat to forests worldwide. In recent years, new diseases have been introduced to Europe and North America at a bewildering rate. European ash trees are being killed by the ash borer beetle and the fungus *Chalara*. This disease, a minor problem of Manchurian ash, was inadvertently transported into Russia by Soviet foresters along with this tree species, where it quickly spread to the native ash of Europe, which it is currently devastating. In America, chestnuts were wiped out a hundred years ago by chestnut blight imported with trees from Japan. And today *Armillaria* root rot is currently spreading fast and threatening hundreds of species of trees right across the globe, killing both important conifers such as spruces and hardwoods such as *Eucalyptus*. Indeed the only trees that seem to have resistance to *Armillaria* are larches and birches!

A final problem is that plantation forestry simply does not fit into the short time scale of the modern industrial world. It is difficult to predict how well a new stand of trees will grow over the next fifty years, and it is impossible to predict the future price of the timber they will eventually produce; foresters can't even predict whether anyone will want the timber. So plenty of forests out there contain trees that have proved a disastrous investment. In Britain, most of the larch trees planted after World War II have grown so crooked in our high winds that their timber is essentially worthless, and many plantations have succumbed to a new invasive *Phytophthora* disease. Also in Britain, Sitka spruces have grown well enough to be made into the sturdy mine props they were planted to produce; unfortunately, there are no longer any active coal mines in the country, or pits for them to prop up. And all over Europe there are fine plantations of overmature oaks, planted in the early years of the nineteenth century to ensure timber supplies for shipbuilding. The only problem is that ships have been built from iron and steel for over 150 years!

The world is poised at a difficult time for forests. There are still over 3 trillion trees on the planet, covering over 30 percent of the globe. Armed with more knowledge about our historical and current impact on trees, we must ensure that we do not let the industrial complex of multinational companies make even further inroads into our inheritance or let foresters damage any more of our remaining woodlands. As we shall see in the final chapter, a vital step in that campaign is to mend our broken relationship with trees and forests and with the wood that they produce.

CHAPTER 15

Mending Our
Strained Relationship

In this book I hope I've been able to show just how important our relationship with wood has been to our success as
a species. How it allowed us to come down from the trees and
become a top predator; how it's allowed us to colonize all the
continents bar Antarctica; and ultimately how it enabled us to
monopolize the land for ourselves. In doing so we have transformed the planet; we have cleared vast areas of forest and
altered the composition of the remaining woodland, but until
the last few hundred years we did this in ways that were essentially sustainable. It's obvious, though, that nowadays this is no
longer the case.

In hindsight the fault line in history probably occurred around
1600, when we started to supplement the energy we obtained
from firewood and charcoal—and gradually replaced them—
with fossil fuels. This process stimulated urbanization, the birth
of science, and the rise of capitalism and industrialization. We
used the new and apparently inexhaustible energy resources
provided by coal, gas, and oil to live together in greater numbers, to develop scientific institutions to better understand the

world, and to combine our new knowledge and power to mass-produce new materials and manufactured goods. The result has been four hundred years of unprecedented material progress and economic growth. Never have we been able to support such a large population, live such long and healthy lives, and produce and consume so much. Abandoning our old reliance on wood seems to have been a great idea.

But it hasn't all been positive. Whenever we are liberated from constraints, we are always apt to let our desires run away with us, to indulge ourselves to excess, and to hurt ourselves and the people and places around us. In our continual striving to increase our material comfort and stimulate our senses with novelties, we have multiplied our energy use twentyfold in the last two hundred years. And since the vast majority of the increase was provided by burning fossil fuels, this has raised the carbon dioxide levels in the atmosphere from 280 ppm to over 400 ppm, causing rapid climate change and putting all our futures in peril. Industrialization has also spread damaging forestry practices across the globe, putting the world's forests at further risk. And perhaps most pernicious of all, industrialization has cut us off from nature and disrupted our species-long relationship with wood, and with the trees that produce it. Gone is the everyday familiarity of our hunter-gatherer ancestors with making wooden tools for the hunt and gathering wood for the fire; gone the knowledge of woodsmanship built up by early farmers, growing trees and shaping them for our use; gone the hands-on experience of generations of carpenters, shaping wood and joining it together to construct our houses and furniture. They have all been discarded as so much old-fashioned lumber by four centuries of breakneck industrialization. We live in a high-tech world, surrounded by toys of all kinds and served by electronic gadgets, but increasingly we lack our ancestors' ability to make things for ourselves.

This was most forcefully brought home to me by the experience of my research assistant Mitch Crook in the jungles of Borneo. In our research on the mechanical role of root buttresses, Mitch needed to cut down smallish trees, around ten feet from the ground, so he could then use a sling to pull them over with our small hand winch. But neither he nor his young town-raised field assistant, Emran, could think of a way of getting high enough up the trunks to saw through them. They had to bring in an older guide from the research center to help. Aged around forty, Sabran was one of the indigenous rain forest people and had been brought up in a local village. He quickly summed the situation up, felled a liana, and wound lengths of it around the trees to build improvised ladders. He had plainly learned and retained the ability to work practically with trees and had a feel for the mechanics of wood. His practical woodworking skills far surpassed those of Mitch, who was an extremely talented hands-on researcher, who had built and maintained complex electronic equipment and had huge experience in performing mechanical tests.

Our estrangement from the practical worlds of forestry and carpentry and our resulting mechanical incompetence have impoverished our lives in several ways and made us more insecure and unhappy, as generally occurs whenever one of our relationships breaks down. Psychologists are now starting to quantify the benefits of our relationship with trees and wood, and to prove that they are as important to us as in our hearts we know they must be. Their studies are showing that people benefit both from being in woodland and even more from working within it, planting tree seedlings and cutting coppice, for instance. Burning wood in open fires and wood-burning stoves provide a natural focus for a quiet evening's contentment, a key ingredient of that most Danish of qualities—hygge. People also feel calmer and happier when they are working with wood, crafting artifacts from timber—my editor, for instance, who, in between correcting manuscripts and writing his own thrillers, makes boxes from the lumber he finds on his travels. And my eldest brother, who releases the tensions built up from researching the tree diseases that are blighting the globe by designing and building model yachts. Completing practical projects such as these brings us a feeling of quiet satisfaction that can't be obtained from more passive entertainments. People even benefit from being surrounded by wood and living within wooden buildings; schoolchildren are less disruptive and perform better academically if they are taught in timber-walled classrooms rather than ones with painted or concrete walls.

Having lost these benefits in our everyday lives, therefore, in many ways we are living more impoverished lives than those of our ancestors. And at the same time we are destroying the planet just as surely as we are making ourselves more miserable. But how can we reverse the process? How can we use wood to heal our planet and bring meaning back to our lives?

There is no doubt that science and technology can play an important role. As we saw in chapter 13, modern ways of transforming timber are allowing it to replace energy-costly materials such as steel and concrete. The new skyscrapers and apartment buildings constructed from glulam and CFL are up to five times lighter than ones made of conventional reinforced concrete. This means they use less energy to build and need shallower foundations, so that the amount of embodied energy they contain can be as low as 20 percent that in a normal building. Given that concrete accounts for around 5 percent of global carbon emissions and steel 3 percent, building a future world with wood could make a significant contribution to countering climate change—around 2.3 percent of the total, a recent report by the Royal Society has estimated. And around the world, wood-science laboratories are coming up with innovations all the time that could enable wood to replace other materials. Professor Liangbing Hu at the University of Maryland has shown that the low stiffness of wood can be overcome by densifying it. Part of the lignin is first removed using a process like that used in papermaking, and the wood is then compressed at high temperature until it is solid. The cellulose fibers reorient themselves along the grain, making the new material not only twelve times stronger than timber but three times as tough; it can be used as a replacement for steel or aluminum. Scientists at the Wallenberg Wood Science Center have shown that modified wood can even replace glass. Removing all the lignin makes it colorless, and the scientists then impregnate the material with a resin with the same refractive index as the cell walls, making the material transparent. Methods of making biodegradable plastics from wood are also well under development across Finland. The way seems to be open to develop a whole new low-carbon economy based on these novel wood products.

However, there are problems with this high-tech approach. For a start, none of these wood products are carbon neutral. Energy is needed to harvest, transport, and machine the wood, for instance. And the most energy-intensive step in all modern wood production is kiln drying. The energy to evaporate water is around 1 megajoule per pound (MJ/lb), and newly felled wood can contain so much of it that kiln drying makes up the bulk of the embodied energy in all wood products, around 4.5 MJ/lb of dry wood. All this can involve the production of greenhouse gases, unless the energy is supplied using renewable sources, or by burning waste wood. Another potential downside to the technological approach is that it would further increase demand for timber, which would drive the expansion of the sort of damaging logging and plantation forestry operations we examined in the last chapter. And the technological fix would continue to move our juggernaut economy even further along the path of unlimited economic growth, which would cause more environmental damage. Forestry already has to compete with farming and nature conservation for land; increasing industrial timber production would reduce food security and biodiversity. Treating timber as yet another industrial product also does nothing to heal our psychological wounds and help rebuild our personal relationship with wood. Living in timber buildings might make us a bit happier, but we would still not have regained our connection to the forest or our woodworking skills.

At the other end of the scale, we could do a great deal locally to rebuild our relationship. Those of us who live in cities are already benefiting from the many urban forestry initiatives that have sprung up around the world, and which are regreening these most artificial environments. Trees have been grown in cities from as long ago as 1840, when the first urban park, Derby Arboretum, was established by Joseph Strutt, the industrialist and brother of the pioneer of iron-framed buildings; by 1858 the best-known urban park, New York's Central Park, was being laid out in Manhattan. Pleached lime trees were a key feature of Georges-Eugène Haussmann's Parisian boulevards, while pollarded plane trees planted in Victorian times still grace the streets of London. More recently city planners have sought to add to these historic legacies. The Million Tree Initiatives have been greening the streets of Los Angeles and New York for the last fifteen years and have stimulated the launch of similar schemes in cities around the world. And for the last thirty years, scientists, economists, and psychologists have sought to quantify the benefits of urban trees, led by the Chicago Urban Forest Project, which was set up by the US Department of Agriculture Forest Service. The research is showing that the benefits of urban trees are considerable. They provide shade, reducing people's effective temperature by 18°F–27°F, and the urban heat island by 2°F–4°F, and can lower air-conditioning costs by 15–30 percent in hot weather. They capture soot particles, lower particulate pollution by around 15 percent, reduce storm runoff by on average 20 percent, and block urban noise. They raise property prices and reduce vandalism. They even make people living on tree-lined streets feel happier and more connected to their neighborhoods, even if they don't actually like trees!

But though people benefit from urban trees when they are growing, the same is not true when the trees are cut down at the end of their lives. Once they are felled, arborists simply shred urban trees into wood chips and sawdust, which are at best used as a soil mulch, so the potential benefits not only of storing carbon but also working and using the wood are lost. Consequently, urban forestry fails to educate city dwellers about our practical relationship with trees; people continue to see them as beautiful, but essentially useless, organisms, rather than as partners in life, and potential sources of firewood and timber. One simple approach to overcome this omission would be to plant larger areas of waste ground with trees such as hazel or willow, which could be managed as coppice. The trees would grow rapidly, producing an area rich in wildlife, while providing an excellent contribution toward cooling the city and preventing flooding. And the areas could be cut in a regular rotation, just like traditional coppice. Volunteers could harvest the shoots for firewood and for woodland crafts such as basket weaving and green woodworking.

Urban foresters could also use portable sawmills to process the timber from larger street trees, providing the basis for further woodworking projects. Similar sorts of urban forestry initiatives have already been tried in Bangladesh, where city dwellers are paid to look after young teak trees that are planted in their neighborhood, with a final payment being made to them when the mature trees are harvested after around twenty years. And the Green Belt Movement is another success story using a bottom-up approach. Set up in Kenya in 1977 by the late Nobel Prize–winner Wangari Maathai, the movement uses a grassroots approach to enable women to grow and plant seedling trees around their homes; it enables them to provide firewood, timber, and food for their families, while binding the soil and storing rainwater, and, most important of all, empowering them economically.

These urban initiatives are on too small a scale to affect wood production globally or even regionally, so they need to be supplemented by larger-scale local or regional efforts. In recent years Ethiopia has had a particularly ambitious reforestation program to plant almost 15 million acres of rural land with trees. In 2019 it harnessed local people to plant 350 million trees in just one day. But we can learn a lot about how to reforest the planet from the regions where the people continue to have a practical, hands-on relationship to the trees that surround them: in the conifer forests of the Pacific Northwest, in the Appalachian Mountains, and in the Alpine and Scandinavian countries of Europe. In these areas, the traditional forestry operations have long tended to be friendly to their environment. And being aware of the fragility of the thin soils on which their conifers grow, foresters are increasingly returning to the methods of their forebears. Rather than clear-felling their forests, they often operate what is known as continuous cover forestry: they take out small numbers of trees at a time, reducing soil damage and allowing the forest to recover faster between fellings, and allowing seedlings to grow under the shelter of mature trees. This produces healthier forests, more pristine landscapes, and better, straighter-grained timber. This approach is being championed by international NGOs such as the Forest Stewardship Council and the Sustainable Forestry Initiative who certify wood that they consider to have come from sustainable woodlands. So far, most of this has come from Scandinavia and North America, but as governance in developing countries improves, there is evidence that their forestry policies do also; the corrupt sale of logging concessions gives way to more sustainable forms of forest management. Various techniques are being trialed. In low-impact logging, vines are cut away and individual trees are carefully pulled out along special tracks, just as colonial foresters used to remove timbers using elephants. In enrichment planting, logged areas are planted with seedlings of valuable canopy trees. Together these techniques reduce environmental damage and speed up the recovery of tropical rain forests and can help tropical timbers receive certification.

But in Scandinavia it is not only the foresters who have a relationship with the trees. In his recent book *Norwegian Wood*, Lars Mytting waxed lyrical about the importance to all Norwegians of the culture of cutting, splitting, storing, and burning firewood. And of all European countries, the one with the strongest forest culture is Finland, where conifer forests still cover over 75 percent of the land. Back in 2001 I was invited to give a couple of talks on trees to accompany a traveling Finnish exhibition, *The Forest and Me*, being hosted at Manchester's Science and Industry Museum. The exhibition outlined everything one might care to know about the forest industry. There were amazing tree-harvesting machines; information about wood pulping; an opportunity to run your own virtual-reality sawmill. Clearly the Finns see the forest as central to their lifestyle and identity; for young Finns the exhibition would have been an excellent introduction to the adult world, to the industries that make their wooden houses and provide their fuel, and to the jobs to which they might aspire. Unfortunately, to an average Mancunian, more interested in the songs of Take That, or to the exploits of Manchester United, it would have seemed irrelevant. Manchester is a big city, situated within rich farmland where the broad-leaved woodland was cleared thousands of years ago, and where people had been burning coal rather than wood for centuries. The exhibition was not a success and attendance was low.

But even in places where the majority of land has been cleared for agriculture (and even near Manchester), there are still numerous small areas of broad-leaved woodland that have been left largely unmanaged over the last century, especially since the invention of plastics made many wood products obsolete. With people hidebound by the assumptions of large-scale plantation forestry, they have been considered too small to be economic, and so they have been left to grow untended and unloved; nowadays their canopies create a dense shade that is killing off the wildflowers that used to flourish beneath them. Fortunately, though, small-scale cooperative groups and some larger companies are at last starting to open up these wood-lands, managing them for firewood, timber, and charcoal, just as our ancestors did hundreds of years ago. The firewood can be sold to local homeowners, who are acquiring wood-burning stoves in ever-increasing numbers, while the timber can be sawn up in local sawmills and sold to craftspeople. There is a rapid expansion of green woodworking, carpentry, and wood turn-ing that is producing furniture, oak buildings, and all manner of the useful tools and items that would have been familiar to our ancestors. The woodlands are starting to act once again as the foundations of a small-scale circular economy.

Meanwhile, the rewilding movement is starting to reclaim large areas of marginal farmland for natural forests and scrub. Trials are showing that this can have huge benefits even in the heavily modified countryside of Britain. On a relatively small scale, discontinuing plowing of heavy clay soils in lowland areas, as at the Knepp Estate in Sussex, England, has allowed the regrowth of scrub and deciduous woodland, while stocking the land with low densities of cattle and pigs is re-creating the wood pasture of medieval times. On the grassy uplands, long degraded by centuries of sheep grazing, removal of these ani-mals in areas such as the Southern Uplands and the Highlands of Scotland are allowing trees to flourish again on what were for-merly green deserts. There are far greater opportunities in con-tinental Europe, where huge areas of marginal land are being abandoned as the young move to the cities. By 2030 this could amount to almost 120,000 square miles of regenerating for-est. And even greater areas are already being rewilded in North America, the largest being the Yellowstone to Yukon Conserva-tion Initiative, which seeks to rewild a strip almost two thou-sand miles long and forty miles wide, an area of around half a million square miles. And across the world, Allan Savory, the Zimbabwean ecologist, has estimated that 19 million square miles of degraded grassland could be restored. The new areas of woodland and scrub that spring up in rewilded sites not only promote high levels of biodiversity, hosting wildflowers, insects, and birds, but they also absorb carbon dioxide and help reverse climate change, just as has been happening over the last century on the abandoned farms of New England and New Zealand. It has been estimated that these vast areas could reabsorb billions of tons of carbon, both in trees and in the soil beneath them, and so lower the CO_2 level by up to 20 ppm, helping to keep global warming down to the manageable level of 2.7°F–3.6°F.

Few of us own any land that we can allow to revert to forest, but each of us can make a difference, first of all by learning more about trees and woodland, and allowing our children to do the same. There is already a flourishing network of Forest Schools around the world, which encourage children to play and learn in natural surroundings. But children get enough input from adults already. Why not let them play in the woods on their own again and teach themselves about their delights and dangers, about the different types of trees and the mechanics of wood? Why not take them to open-air museums? At the very least they can run around a lot in the fresh air, get rid of their energy, and enjoy themselves far more than they would in a traditional museum; and they are more likely to learn interesting things about the everyday life of our ancestors. Why not teach them how to identify trees and search for wild food? Or how to use their hands making wooden models or building useful things? I'm no expert, but I managed to make insect cages and nest boxes when I was a kid, and even a "modern sculpture," which I carved in a woodworking lesson and which is still performing a useful service at my father's house, acting as a kitchen paper towel holder. These may not have been perfect or even well made, but they meant a lot more to me than if I'd just bought them. So why not abandon our continual acquisition of more and more manufactured goods and buy, or preferably learn to make, just a few simple wooden things for ourselves? You need not go whole hog and live in a Neolithic-style round-house, like one of my former colleagues, or teach your kids to carve bows and arrows, like my former PhD student Adrian Goodman. But any reduction in the amount of goods we buy can reduce our impact on the planet. Who knows, we might be able to start a ball rolling and help the human race return to the more gentle delights of the Age of Wood.

Acknowledgments

This book is the fruit of many years of wandering and pondering, most of all stimulated by my role as an academic at the Universities of Manchester and Hull. This gave me the privileges of being able to travel to forests around the world, ostensibly for teaching and research, and to teach bright students about subjects as varied as biomechanics, evolution, plant biology, and trees. It gave me free rein to try and fit together all the information I was gleaning. I thank my project and research students for testing out many of my wild ideas, particularly the students mentioned by name throughout the book. Without their talent and enthusiasm I would never have been able to build up my story.

I thank my agent, Peter Tallack, for taking me on, and helping shape my proposal into a coherent story, and rid it of much of my academic obfuscation. I thank my editors at Simon & Schuster, Colin Harrison and Sarah Goldberg, for helping to shape the book and teaching me something of American history. I thank the people who so kindly read and commented on sections of the book: my brother Richard Ennos, my friends and colleagues Peter Lucas, Adam van Casteren and David Armson, and the selfless Lindsay Wood.

I thank my family for setting me on the lifelong trail of finding connections and trying to make sense of the world around me. I thank Hampton Grammar School (now Hampton School), for giving me the confidence, attitude, and intellectual tools to challenge conventional wisdom. And most of all thank you to my partner, Yvonne, for thirty years of support, love, companionship, and for teaching me the many benefits of pottering.

Notes

Most of the information set out in this book, particularly historical facts and statistics, is freely available online, through websites such as Wikipedia. However, facts are just the building blocks of real knowledge and understanding. The references set out below present more useful information in the form of original research articles, reviews, and books that have aimed to link such information to tell stories about what we know, how we know it, and why we believe it is true.

Prologue: The Road to Nowhere

xii *the King's Broad Arrow policy*: See Malone (1979) for a full account.
xii *The situation reached a crisis in 1772*: For a description of the Pine Tree Riot see Danver (2011), 183–90.

Chapter 1: Our Arboreal Inheritance

3 *"Let us make man in our image"*: Genesis 1:26.
5 *This increases friction because a soft material*: See Ennos (2012) and Warman and Ennos (2009) for an account of friction and the design of finger pads.
6 *pattern of ridges known as fingerprints*: For our study see Warman and Ennos (2009).
7 *instead these were flattened into self-trimming nails*: For a description of the ingenious mechanical design of fingernails see Farren et al. (2004).
7 *Bush babies have brains that are only slightly larger*: See Stephan et al. (1981).
10 *macaques and spider monkeys have brains that are on average about 25 percent bigger*: For a recent paper see DeCasien et al. (2017).

10 *this "social hypothesis" does not explain*: Robin Dunbar has written extensively about it in, for instance, Dunbar (2009) and (2016).
12 *"clambering hypothesis" of Daniel Povinelli and John Cant*: See Povinelli and Cant (1995).
13 *they move quite differently when they travel along branches*: See Thorpe et al. (2007a).
14 *They can even exploit the flexibility of tree trunks*: See Thorpe et al. (2007b).
14 *apes have more frequent bouts of both NREM*: See Samson and Shumaker (2015) and Samson and Nunn (2015).
14 *Wood is quite a complex material*: For details about wood anatomy see Ennos (2016).
15 *This complex structure gives wood different mechanical properties*: For discussions about wood properties and greenstick fracture see Ennos and van Casteren (2010) and van Casteren et al. (2012a).
17 *What Adam found*: The investigation is described in van Casteren et al. (2012b), while the last part of Julia's video seems to be on You Tube at https://www.youtube.com/watch?v=g6gfG4aCUyw.
18 *orangutans make rather few tools in the wild*: See van Schaik (2004) and van Schaik and Knott (2001).
19 *The honey-loving chimps of Gabon are even more sophisticated*: See Boesch et al. (2009).
19 *The savanna chimps of Tanzania, East Africa*: See Hernandez-Aguilar et al. (2007).
19 *Jill Pruetz of Texas State University has observed*: See Pruetz and Bertolani (2007).
20 *for an alternative hypothesis*: See Thorpe et al. (2007a).
21 *He measured the stiffness*: See van Casteren et al. (2013).
21 *holding on to branches could help an animal overcome*: See Johannsen et al. (2017).
22 Orrorin tugenensis, *one of the earliest*: See Henke et al. (2007).
22 *The hip and leg bones of the 4.4-million-year-old* Ardipithecus ramidus: See Lovejoy et al. (2009).
22 *The 12-million-year-old fossil ape* Danuvius guggenmosi: See Böhme et al. (2019).

Chapter 2: Coming Down from the Trees

25 *Lucy, had met a violent death falling from a tall tree*: See Kappelman et al. (2016).
25 *Bill Sellers of the University of Manchester*: See Sellers et al. (2005).
26 *footprints left in sand by even earlier australopiths*: See Crompton et al. (2011).

26 *John Kappelman and his coworkers*: See Kappelman et al. (2016).
27 *showed that she had apelike shoulder blades*: See Green and Alemseged (2012).
27 *CT scans of Lucy's bones*: See Ruff et al. (2016).
27 *this youngster had even more curved metatarsal joints than Lucy*: See DeSilva et al. (2018).
27 *only with the emergence of* Homo erectus: See Wrangham (2009).
28 *The exposure of the forest floor to light*: For more on plant adaptations to drought and plant roots see Ennos and Sheffield (2000).
29 *Both the early australopiths and* Homo habilis *developed their dentition*: For more on adaptations of early hominids to diet see, for instance, Leakey (1996).
30 *modern savanna chimps use digging sticks*: See Hernandez-Aguilar et al. (2007).
30 *the felling of taprooted plants*: For a summary of this research on anchorage see Ennos (2000).
30 *the earliest digging sticks that have been found*: See Aranguren et al. (2018).
31 *The digging sticks used by modern-day hunter-gatherers*: For details of these see Vincent (1984).
32 *the molecular structure of the cell walls themselves*: For more on the mechanics of wood see Gordon (1968).
33 *Their brains were not much larger than those*: For more on the brain size of hominins see Wrangham (2009).
34 *they get a rotten night's rest*: See Samson and Shumaker (2015) and Samson and Nunn (2015).
35 *The evaporation of water within the cells is what destroys trees*: This was first demonstrated in the nineteenth century by Osborne Reynolds, the founder of the science of hydrodynamics.
39 *The consequences of the mechanical breakdown of food by cooking*: See Wrangham (2009), chap. 6.
40 *But the chemical breakdown of food that occurs*: See Wrangham (2009), chap. 3.
41 *The fossil evidence . . . points to an early date of over 1.6 million years ago*: See Wrangham (2009), chap. 4.
42 *there is less direct evidence in the form of the remains of fires*: See Wrangham (2009), chap. 4, and Gowlett (2016).

Chapter 3: Losing Our Hair

43 *Desmond Morris's popular 1960s take*: See Morris (1967).
43 *However, modern molecular genetics has been able to cast some light*: See Rogers et al. (2004).

44 *This idea has been widely expounded from the 1960s*: See, for instance, Wheeler (1992).
45 *losing our hair was crucial for another advance*: See Morris (1967).
45 *fine BBC video of a San Bushman tracking*: See Attenborough (2009).
46 *In the heat of the day a naked body would actually absorb* more *heat*: See Queiroz do Amaral (1996), a rare paper by a woman on the subject, which has been widely ignored by men!
46 *the net radiation . . . can amount to around 670 watts per square yard*: See Ruxton and Wilkinson (2011).
48 *several scientists have championed an alternative hypothesis*: See Rantala (2007) for a review of the field.
49 *recent research by Isabelle Dean and Michael Siva-Jothy*: See Dean and Siva-Jothy (2012).
51 *the first actual physical evidence for clothes*: See Rantala (2007).
51 *Sumatran orangutans . . . often make second nests directly above*: See van Schaik (2004).
51 *Indeed, many tribes of hunter-gatherers still build small semipermanent huts*: See for instance Turnbull (1961) and Samson et al. (2017).
51 *evidence of a 1.8-million-year-old building*: See Leakey (1971).
52 *Research on the climatic benefits*: See Armson et al. (2012).
52 *Samson and his team also estimated how warm people would feel*: See Samson et al. (2017).

Chapter 4: Tooling Up

56 *In his 1865 book,* Prehistoric Times, *Lubbock*: See Lubbock (1865).
58 *For instance in his otherwise admirably clear book*: See Leakey (1996).
62 *Lawrence Keeley and Nicholas Toth of the University of Illinois*: See Keeley and Toth (1981).
62 *Manuel Dominguez-Rodrigo and coworkers*: See Dominguez-Rodrigo et al. (2001).
62 *Miriam Haidle, from the University of Tübingen*: See Haidle (2009).
64 *The earliest recorded wooden tool is the Clacton Spear*: See Warren (1911).
64 *Modern experiments to replicate how it was made*: See McNabb (1989) and Fluck (2007).
65 *I set an undergraduate project student, Michael Chan*: For the results see Ennos and Chan (2016).
65 *Thieme and his colleagues were astonished to uncover*: For the discovery of the spears see Thieme (1997), and for subsequent analysis see a series of papers in *Journal of Human Evolution*, Conard et al. (2015).

66 *experiments on replicas have shown*: See Milks et al. (2019).
67 *Many anthropologists have carried out experimental investigations*: See, for instance, Waguespack et al. (2009).
68 *the unlikely pursuit of arrow throwing*: See Westcott (1999), 192–94.
70 *to propel a small spear or dart*: See Westcott (1999), 195–99.
70 *Even better results can be achieved*: See Westcott (1999), 200–209.
71 *They invented a wide range of boomerangs*: See Westcott (1999), 210–24.
72 *Marlize Lombard and Miriam Haidle have calculated*: See Lombard and Haidle (2012).
73 *we had used wooden tools to kill off*: See Kolbert (2014).

Chapter 5: Clearing the Forest

77 *a major recent account of the Neolithic in Europe*: See Miles (2016).
78 *They could have cut down small saplings*: See Fowler (1962).
79 *the Dalton people of the Mississippi also produced*: See Yerkes and Koldehoff (2018).
79 *These were the remains of a circular hut*: See Waddington (2007).
80 *Remains of an even earlier hut*: See Milner et al. (2013).
80 *However, in 2007 a yard-long piece of oak*: See Momber et al. (2011).
81 *the first watercraft that were developed*: For more on these skin boats see Elmers (1996).
82 *A quite different sort of boat*: For more on log boats see Elmers (1996).
86 *Svend Jørgensen and his colleagues showed*: See Jørgensen (1985).
86 *I decided to investigate their design*: For the results see Ennos and Oliveira (2017).
87 *Modern splitting mauls also have broad, heavy heads*: See Mytting (2015).
87 *That it was also a problem for Neolithic people*: See Taylor (1998).
87 *But Neolithic people seem mostly to have been adept*: See the reconstruction by Harding (2014).
88 *In the axes found at the Neolithic lakeside villages*: See Bugrov and Galimova (2017).
89 *Native Americans bound the joint up*: See Fowler (1962).
89 *Recent experiments by Rengert Elburg and his group*: See Elburg et al. (2015).
89 *Duncan showed that within the crotch of these joints*: See Slater et al. (2014).
89 *Rengert Elburg of the Archaeological Heritage Office*: See Elburg et al. (2015).

90 *Phil Harding . . . has been able to replicate*: See Harding (2014).
91 *The most impressive achievement of the LBK people*: See Miles (2016).
92 *Willy Tegel from the University of Freiburg, Germany*: See Tegel et al. (2012).
93 *the excavations of the stone-walled houses*: For more about the houses at Skara Brae and Durrington Walls see Miles (2016).
94 *this allows the shoots to grow straighter, stiffer*: See Ozden and Ennos (2018).
95 *The Sweet Track, as it became known*: See Coles and Coles (1988).
96 *People still make hurdles*: For full instructions see Law (2015).
96 *Neolithic farmers made small lightweight rounded boats*: See Elmers (1996).
97 *though farming emerged at different times*: See Diamond (1997).

Chapter 6: Melting and Smelting

101 *the first-known clay sculpture is the Venus of Dolní Věstonice*: See Vandiver et al. (1989).
101 *first fired bricks were made about 4300 BC*: See Yasuda (2012).
101 *the first roof tiles replaced thatch in Mesopotamia*: See Woods and Woods (2000).
102 *Charcoal-fired kilns also enabled craftsmen*: See Whitehouse (2012).
103 *copper combines some of the best material properties*: See Gordon (1968).
105 *The ice man Ötzi*: See Fleckinger (2018).
106 *James Mathieu . . . has shown*: See Mathieu and Mayer (1997).
107 *many Bronze Age boats have started to be discovered*: For a discussion of Bronze Age boats see McGrail (1996).
108 *The funeral ship of Khufu*: See Jenkins (1980).
109 *Turkish archaeologists discovered a late Bronze Age vessel*: See Pulak (1998).
110 *Dick Parry has suggested that the Egyptians*: See Parry (2004).
111 *Consequently when a wooden disk dries out*: See Hoadley (2002).
112 *The first wheels were therefore cut from planks of wood*: For early wheel design see Anthony (2007).
112 *Wheels appeared almost simultaneously in the Sumerian civilization*: See Anthony (2007).
114 *there are two exceptions*: For a review of the origin of plank boats in the New World see Gamble (2002).
115 *It is much-better known that none*: See, for instance, Diamond (1997), who cites the absence of useful draft animals as the reason why wheeled vehicles were never used in the Americas.

Chapter 7: Carving Our Communities

121 *the trunks of trees are actually prestressed*: For more detail see Gordon (1978) and Ennos (2016).

122 *in the Song dynasty, the Chinese developed*: For more details see Zhou et al. (2018).

125 *the great Scandinavian stave churches*: For more on these buildings see Pryce (2005).

127 *the greatest triumph of the green woodworking tradition was*: For more on longship design see Christensen (1996) and Durham (2002).

130 *The first item to be added*: For more on the design and use of traditional carpentry tools see Bealer (1996).

136 *most famous design being the classical Windsor chair*: For more on Windsor chairs see Green (2006).

136 *To make a barrel, coopers had first*: See Logan (2005).

137 *wheel design reached a peak in the chariots of the Homeric Greeks*: For more details see Gordon (1968).

Chapter 8: Supplying Life's Luxuries

140 *Wood is naturally adapted to the needs*: For more details of the adaptations of wood see Ennos (2016).

146 *lines of large vessels along their growth rings*: For more about the reasons why oak wood has such an unusual anatomy see Ennos (2016).

Chapter 9: Supporting Our Pretensions

155 *Geoff Carter has even suggested that the sarsen stones*: See Carter (2012), just one of his many blogs about structural archaeology.

157 *Essentially their architects cheated*: For details about the architecture of the roofs of Greek temples, see Hodge (1960).

158 *The book of Kings, for instance, tells us*: See Kings. 7:23–26.

166 *Recent research by Chinese engineers shows*: See a recent Channel 4 documentary at https://www.channel4.com/programmes/secrets-of-chinas-forbidden-city/episode-guide/.

Chapter 10: Limiting Our Outlook

170 *For instance, in 240 BC the tyrant of Syracuse*: See Casson (1996).
171 *the Romans later built entirely practical ships*: See Casson (1996).
173 *In the Middle Ages, around thirty pounds of wood was needed*: See Wrigley (2010).
173 *in England and Wales in the 1650s, people obtained*: See Warde (2007).
175 *English woodsmen traditionally cut coppiced firewood*: See Rackham (2003).
175 *the town of Odense in Denmark, with a population*: See van der Woude et al. (1990).
176 *Paris . . . obtained the majority of its firewood supplies*: See Jouffroy-Babicot et al. (2013).
181 *diagonal members incorporated between the rafters*: See Yorke (2010).
182 *so bend the ship, a deformation known as hogging*: For more details about hogging and hogging trusses see Gordon (1978).

Chapter 11: Replacing Firewood and Charcoal

187 Fumifugium *he railed against the smogs*: See Evelyn (1661).
189 *Jan de Zeeuw of the Agricultural University of Wageningen has calculated*: See de Zeeuw (1978).
189 *three times the energy per person than the English*: See Wrigley (2010).
193 *Francis Bacon, who, in*: The New Atlantis: See Bacon (1627).
193 *the first-ever set of DIY manuals*: See Moxon (1703).
194 Sylva *sought to bring together all that was known*: See Evelyn (1664).
196 *such as the Lunar Society*: See Uglow (2002) for more details.
197 *it produced only around twenty-five thousand tons of iron*: See Wrigley (2010).
199 *John Wilkinson, developed the first precision cutting tool*: For more on Wilkinson's work see Winchester (2018).
201 *From 3.5 million tons in 1700*: See Pollard (1980).
201 *European rulers also hastened to maintain their wood supply*: For more details of the situation in central Europe see Radkau (2012).
203 *new techniques to improve the quality of charcoal*: For more on the American iron industry see Schallenberg (1981).

Chapter 12: Wood in the Nineteenth Century

215 *To make each block he therefore designed*: For more on the story of Brunel's block-making machines see Winchester (2018).
217 *As early as 1812, for instance, Louis Wernwag*: For more on this story see Nelson (1981).
217 *Americans built their railroads for the most part with timber*: For the background about American railroads see White (1981).
219 *an even simpler and humbler attachment system—nails*: See Rybczynski (2000).
220 *The builders could then nail these uniform elements*: See Green (2006).
221 *the machine-made wood screw, made all sorts of structures*: For more on the story of screws and screwdrivers see Rybczynski (2000).
222 *one nineteenth-century development must be more important*: For more on the history of paper see Kurlansky (2017).
223 *a dramatic increase in the size of newspapers, and their circulation*: See Smith (1964).

Chapter 13: Wood in the Modern World

231 *As oil became available in the twentieth century*: For more on plastics see Gordon (1968).

Chapter 14: Assessing Our Impact

247 *so many people exaggerate the pace of erosion*: See Ennos and Bailey (1995), Problem 5.1.
250 *Take one of the most cited myths*: See Diamond (2011) for the conventional view.
253 *It was the states on former grassland and desert regions*: The information, though not the connection, can be found in Diamond (2011).
257 *the recent finds of Amazonian Dark Earths*: See Heckenberger and Neves (2009).
258 *tree cover fell from a maximum of around 80 percent*: See Roberts et al. (2018).
258 *the forest cover of England*: See Rackham (2006).
259 *the loss of conifer forests in Scandinavia and the Alps was undetectable*: See Roberts et al. (2018).

259 *Thousands of years of woodland management have also drastically*: For more details see Rackham (2003) and (2006).

260 *Michael Williams of the University of Oxford has estimated*: See Williams (2002).

260 *this would have released something in the region of 275 Gtons*: See Ruddiman (2003).

261 *Pollen analysis, for instance, has shown that the forest cover of Europe*: See Roberts et al. (2018).

261 *another major historical event is now thought to have helped cause*: For more on this story see Koch et al. (2019).

262 *our effect on the world's forests has accelerated rapidly*: For more details see Williams (2002).

Chapter 15: Mending Our Strained Relationship

271 *in all wood products, around 4.5 MJ/lb of dry wood*: See Jones (2019).

273 *the benefits of urban trees are considerable*: See my mini-review on the physical benefits of urban trees in Hirons and Thomas (2018).

277 *Meanwhile, the rewilding movement is starting to reclaim*: See Tree (2017).

References

Anthony, D. A. 2007. *The Horse, the Wheel, and Language: How Bronze-Age Riders from the Eurasian Steppes Shaped the Modern World.* Princeton, NJ: Princeton University Press.

Aranguren, B., A. Revedin, N. Amico, F. Cavulli, G. Giachi, S. Grimaldi, N. Macchioni, and F. Santaniello. 2018. "Wooden Tools and Fire Technology in the Early Neanderthal Site of Poggetti Vecchi (Italy)." *Proceedings of the National Academy of Sciences* 115:2054–59.

Armson, D., P. Stringer, and A. R. Ennos. 2012. "The Effect of Tree Shade and Grass on Surface and Globe Temperatures in an Urban Area." *Urban Forestry and Urban Greening* 11:245–55.

Attenborough, D. 2009. *Human Mammal, Human Hunter: Life of Mammals.* BBC.

Bacon, F. 1627. *The New Atlantis.*

Bealer, A. W. 1996. *Old Ways of Working Wood.* Edison, NJ: Castle Books, 1996.

Boesch, C., J. Head, and M. M. Robbins. 2009. "Complex Tool Sets for Honey Extraction among Chimpanzees in Loango National Park, Gabon." *Journal of Human Evolution* 56:560–69.

Böhme, M., N. Spassov, J. Fuss, A. Tröscher, A. S. Deane, J. Prieto, U. Kirscher, T. Lechner, and D. R. Begun. 2019. "A New Miocene Ape and Locomotion in the Ancestor of Great Apes and Humans." *Nature* 575:489–93.

Bugrov, D., and M. Galimova. 2017. "Antler Sleeves from the Neolithic Lake-Dwelling Sites of Switzerland (the 'Swiss Collection' of the National Museum of Tatarstan Republic, Kazan)." *Povolzhskaya Arkheologiya* 1:26–37.

Carter, G. 2012. "Twelve Reasons Why Stonehenge Was a Building." *Theoretical Structural Archaeology*, March 23. http://structuralarchaeology.blogspot.com/2012/03/twelve-reasons-why-stonehenge-was.html.

Casson, L. 1996. "Sailing Ships of the Ancient Mediterranean." In *The Earliest Ships*, edited by R. Gardiner, 39–51. London: Conway Maritime Press.

Christensen, A. E. 1996. "Proto-Viking, Viking and Norse Craft." In *The Earliest Ships*, edited by R. Gardiner, 72–88. London: Conway Maritime Press.

Coles, B., and J. Coles. 1988. *Sweet Track to Glastonbury: Somerset Levels in Prehistory (New Aspects of Antiquity)*. London: Thames and Hudson.

Conard, N. J., C. E. Miller, J. Serangeli, and T. van Kolfschoten. 2015. "Special Issue: Excavations at Schöningen: New Insights into Middle Pleistocene Lifeways in Northern Europe." *Journal of Human Evolution* 89:1–308.

Crompton, R. H., T. C. Pataky, K. Savage, K. D'Aout, M. R. Bennett, M. H. Day, K. Bates, S. Morse, and W. I. Sellers. 2011. "Human-Like External Function of the Foot, and Fully Upright Gait, Confirmed in the 3.66 Million-Year-Old Laetoli Hominin Footprints by Topographic Statistics, Experimental Footprint-Formation and Computer Simulation." *Journal of the Royal Society Interface* 9:707–19.

Danver, S., ed. 2011. *Revolts, Protests, Demonstrations, and Rebellions in American History: An Encyclopedia*. ABC-CLIO, LLC, s.v., "Pine Tree Riot," 183–90.

Dean, I., and M. T. Siva-Jothy. 2012. "Human Fine Body Hair Enhances Ectoparasite Detection." *Biology Letters* 8:358–61.

DeCasien, A. R., S. A. Williams, and J. P. Higham. 2017. "Primate Brain Size Is Predicted by Diet but Not Sociality." *Nature Ecology and Evolution* 1:0112.

DeSilva, J. M., C. M. Gill, T. C. Prang, M. A. Bredella, and Z. Alemseged. 2018. "A Nearly Complete Foot from Dikika, Ethiopia, and Its Implications for the Ontogeny and Function of *Australopithecus afarensis*." *Science Advances* 4:7723.

de Zeeuw, J. W. 1978. "Peat and the Dutch Golden Age. The Historical Meaning of Energy-Attainability." *A.A.G. Bijdragen* 21:3–31.

Diamond, J. 1997. *Guns, Germs and Steel*. London: Jonathan Cape.

———. 2011. *Collapse: How Societies Choose to Fail or Survive*. London: Penguin.

Dominguez-Rodrigo, M., J. Serrallonga, J. Juan-Tresserras, L. Alcala, and L. Luque. 2001. "Woodworking Activities by Early Humans: A Plant Residue Analysis on Acheulian Stone Tools from Peninj (Tanzania)." *Journal of Human Evolution* 40:289–99.

Dunbar, R. 2009. "The Social Brain Hypothesis and Its Implications for Social Evolution." *Annals of Human Biology* 36:562–72.

———. 2016. *Human Evolution: Our Brains and Behavior*. Oxford: Oxford University Press.

Durham, K. 2002. *Viking Longship*. Oxford: Osprey.

Elburg, R., W. Hein, A. Probst, and P. Walter. 2015. "Field Trials in Neolithic Woodworking." In *Archaeology and Crafts—Experiences and Experiments on Traditional Skills and Handicrafts in Archaeological Open-Air Museums in Europe*, edited by R. Kelm, 62–77. Husum, Germany: Husum Druck- und Verlagsgesellschaft.

Elmers, D. 1996. "The Beginnings of Boatbuilding in Central Europe." In *The Earliest Ships*, edited by R. Gardiner, 11–23. London: Conway Maritime Press.

Ennos, A. R. 2000. "The Mechanics of Root Anchorage." *Advances in Botanical Research* 33:133–57.

———. 2012. *Solid Biomechanics*. Princeton, NJ: Princeton University Press.

———. 2016. *Trees*. 2nd ed. London: Natural History Museum and University Press.

Ennos, A. R., and S. E. R. Bailey. 1995. *Problem Solving in Environmental Biology*. Harlow, UK: Longman's Higher Education.

Ennos, A. R., and M. Chan. 2016. "'Fire Hardening' Spear Wood Does Slightly Harden It, but Makes It Much Weaker and More Brittle." *Biology Letters* 12.

Ennos, A. R., and J. A. V. Oliveira. 2017. "The Mechanics of Splitting Wood and the Design of Neolithic Woodworking Tools." Exarc.Net.

Ennos, A. R., and E. Sheffield. 2000. *Plant Life*. Oxford: Blackwell Science.

Ennos, A. R., and A. van Casteren. 2010. "Transverse Stresses and Modes of Failure in Tree Branches and Other Beams." *Proceedings of the Royal Society B* 277:1253–58.

Evelyn, J. 1661. *Fumifugium*. London: His Majesties' Command.

———. 1664. *Sylva, or a Discourse of Forest Trees*. Minneapolis, MN: Filiquarian.

Farren, L., S. Shayler, and A. R. Ennos. 2004. "The Fracture Properties and Mechanical Design of Human Fingernails." *Journal of Experimental Biology* 207:735–41.

Fleckinger, A. 2018. *Ötzi the Iceman: The Full Facts at a Glance*. Czech Republic: Folio Verlagsges. Mbh.

Fluck, H. L. 2007. "Initial Observations from Experiments into the Possible Use of Fire with Stone Tools in the Manufacture of the Clacton Point." *Lithics* 28:15–19.

Fowler, W. S. 1962. "Woodworking: An Important Industry." *Bulletin of the Massachusetts Archaeological Society* 23:29–40.

Gamble, L. H. 2002. "Archaeological Evidence for the Origin of the Plank Canoe in North America." *American Antiquity* 67:301–15.

Gordon, J. E. 1968. *The New Science of Strong Materials, or Why You Don't Fall Through the Floor*. London: Penguin.

———. 1978. *Structures, or Why Things Don't Fall Down*. London: Penguin.

Gowlett, J. A. J. 2016. "The Discovery of Fire by Humans: A Long and Convoluted Process." *Philosophical Transactions of the Royal Society B* 371.

Green, D. J., and Z. Alemseged. 2012. "*Australopithecus afarensis* Scapular Ontogeny, Function, and the Role of Climbing in Human Evolution." *Science* 338 (6106): 514–17.

Green, H. 2006. *Wood*. New York: Viking Penguin.

Haidle, M. 2009. "How to Think a Simple Spear." In *Cognitive Archaeology*, edited by S. de Beaune, F. Coolidge, and T. Wynn, 55–73. Cambridge: Cambridge University Press.

Harding, P. 2014. "Working with Flint Tools: Personal Experience Making a Neolithic Axe Haft." *Lithics* 35:40–53.

Heckenberger, M., and E. G. Neves. 2009. "Amazonian Archaeology." *Annual Review of Anthropology* 38:251–66.

Henke, W., I. Tattersall, and T. Hart. 2007. *Handbook of Paleoanthropology, III: Phylogeny of Hominids*. Berlin: Springer-Verlag.

Hernandez-Aguilar, R. A., J. Moore, and T. R. Pickering. 2007. "Savanna Chimpanzees Use Tools to Harvest the Underground Storage Organs of Plants." *Proceedings of the National Academy of Sciences* 104:19210–13.

Hirons, A. D., and P. A. Thomas. 2018. *Applied Tree Biology*. Oxford: John Wiley.

Hoadley, R. B. 2002. *Understanding Wood*. Newtown, CT: Taunton Press.

Hodge, A. T. 1960. *The Woodwork of Greek Roofs*. Cambridge: Cambridge University Press.

Jenkins, N. 1980. *The Boat Beneath the Pyramid: King Cheops' Royal Ship*. New York: Holt, Rinehart, and Winston.

Johannsen, L., S. R. L. Coward, G. R. Martin, A. M. Wing, A. van Casteren, W. Sellers, A. R. Ennos, R. Crompton, and S. K. S. Thorpe. 2017. "Human Bipedal Instability in Tree Canopy Environments Is Reduced by 'Light Touch' Fingertip Support." *Scientific Reports* 7:1135.

Jones, C. 2019. "Ice Database of Embodied Energy and Carbon." http://www.circularecology.com/embodied-energy-and-carbon-footprint-database.html#.XN_8adhrzIU.

Jørgensen, S. 1985. *Tree-Felling with Original Neolithic Flint Axes in Draved Wood. Report on the Experiments in 1952–1954*. Copenhagen: National Museum of Denmark.

Jouffroy-Babicot, I., B. Vanniere, E. Gauthier, H. Richard, F. Monna, and C. Petit. 2013. "7000 Years of Vegetation History and Land-Use Changes in the Morvan Mountains (France): A Regional Synthesis." *Holocene* 23:1888–902.

Kappelman, J., R. A. Ketcham, S. Pearce, L. Todd, W. Akins, M. W. Colbert, M. Feseha, J. A. Maisano, and A. Witzel. 2016. "Perimortem Fractures in Lucy Suggest Mortality from Fall out of Tall Tree." *Nature* 537:503–7.

Keeley, L., and N. Toth. 1981. "Microwear Polishes on Early Stone Tools from Koobi Fora, Kenya." *Nature* 293:464–65.

Koch, A., C. Brierley, M. M. Maslin, and S. L. Lewis. 2019. "Earth System Impacts of the European Arrival and Great Dying in the Americas after 1492." *Quaternary Science Reviews* 207:13–36.

Kolbert, E. 2014. *The Sixth Extinction: An Unnatural History.* London: Bloomsbury.

Kurlansky, M. 2017. *Paper: Paging Through History.* New York: W. W. Norton.

Law, B. 2015. *Woodland Craft.* Lewes, UK: Guild of Master Craftsmen.

Leakey, M. D. 1971. *Olduvai Gorge, III: Excavations in Beds I and II, 1960–1963.* Cambridge: Cambridge University Press.

Leakey, R. 1996. *The Origin of Humankind.* New York: Basic Books.

Logan, W. B. 2005. *Oak: The Frame of Civilization.* New York: W. W. Norton.

Lombard, M., and M. H. Haidle. 2012. "Thinking a Bow-and-Arrow Set: Cognitive Implications of Middle Stone Age Bow and Stone-Tipped Arrow Technology." *Archaeological Journal* 22:237–64.

Lovejoy, C. O., G. Suwa, L. Spurlock, B. Asfaw, and T. D. White. 2009. "Pelvis and Femur of *Ardipithecus ramidus*: The Emergence of Upright Walking." *Science* 326:71.

Lubbock, J. 1865. *Prehistoric Times.* London: Williams and Norgate.

Malone, J. J. 1979. *Pine Trees and Politics.* New York: Arno Press.

Mathieu, J. R., and D. A. Meyer. 1997. "Comparing Axe Heads of Stone, Bronze, and Steel: Studies in Experimental Archaeology." *Journal of Field Archaeology* 24:333–51.

McGrail, S. 1996. "The Bronze Age in Europe." In *The Earliest Ships*, edited by R. Gardiner, 24–38. London: Conway Maritime Press.

McNabb, J. 1989. "Sticks and Stones: A Possible Experimental Solution to the Question of How the Clacton Spear Point Was Made." *Proceedings of the Prehistoric Society* 55:251–71.

Miles, D. 2016. *The Tale of the Axe: How the Neolithic Revolution Transformed Britain.* London: Thames and Hudson.

Milks, A., D. Parker, and M. Pope. 2019. "External Ballistics of Pleistocene Hand-Thrown Spears: Experimental Performance Data and Implications for Human Evolution." *Scientific Reports* 25:820.

Milner, N., B. Taylor, C. Conneller, and T. Schadla-Hall. 2013. *Star Carr: Life in Britain after the Ice Age.* York, UK: Council for British Archaeology.

Momber, G., D. J. Tomalin, R. G. Scaife, and J. Satchell. 2011. *Mesolithic Occupation at Bouldnor Cliff and the Submerged Prehistoric Landscapes of the Solent.* York, UK: Council for British Archaeology.

REFERENCES

Morris, D. 1967. *The Naked Ape.* London: Jonathan Cape.

Moxon, J. 1703. *Mechanick Exercises or the Doctrine of Handy-Works.* Wilmington, NC: Toolemera Press.

Mytting, L. 2015. *Norwegian Wood.* London: MacLehose Press.

Nelson, L. H. 1981. "The Colossus of Philadelphia." In *Material Culture of the Wooden Age*, edited by B. Hindle, 159–83. Tarrytown, NY: Sleepy Hollow Press.

Ozden, S., and A. R. Ennos. 2018. "The Mechanics and Morphology of Branch and Coppice Stems in Three Temperate Tree Species." *Trees* 32:933–49.

Parry, D. 2004. *Engineering the Pyramids.* Stroud, UK: Sutton Publishing.

Pollard, S. 1980. "A New Estimate of British Coal Production, 1750–1850." *Economic History Review* 33:212–35.

Povinelli, D. J., and J. G. H. Cant. 1995. "Arboreal Clambering and the Evolution of Self-Conception." *Quarterly Review of Biology* 70:393–421.

Pruetz, J. D., and P. Bertolani. 2007. "Savanna Chimpanzees, *Pan troglodytes verus*, Hunt with Tools." *Current Biology* 17:412–17.

Pryce, W. 2005. *Architecture in Wood.* London: Thames and Hudson.

Pulak, C. 1998. "The Uluburun Shipwreck: An Overview." *International Journal of Nautical Archaeology* 27:188–224.

Queiroz do Amaral, L. 1996. "Loss of Body Hair, Bipedality and Thermoregulation. Comments on Recent Papers in the *Journal of Human Evolution.*" *Journal of Human Evolution* 30:357–66.

Rackham, O. 2003. *Ancient Woodland.* 2nd ed. Dalbeattie, UK: Castlepoint Press.

———. 2006. *Woodlands.* London: Harper Collins.

Radkau, J. 2012. *Wood: A History.* Cambridge: Polity Press.

Rantala, M. J. 2007. "Evolution of Nakedness in *Homo sapiens.*" *Journal of Zoology* 273:1987–89.

Roberts, N., R. M. Fyfe, J. Woodbridge, M.-J. Gaillard, B. A. S. Davis, J. O. Kaplan, L. Marquer, F. Mazier, A. B. Nielsen, S. Sugita, A.-K. Trondman, and M. Leydet. 2018. "Europe's Lost Forests: A Pollen-Based Synthesis for the Last 11,000 Years." *Scientific Reports* 8:716.

Rogers, A. R., D. Iltis, and S. Wooding. 2004. "Genetic Variation at the MC1R Locus and the Time Since Loss of Human Body Hair." *Current Anthropology* 45:105–8.

Ruddiman, W. F. 2003. "The Anthropogenic Greenhouse Era Began Thousands of Years Ago." *Climatic Change* 61:261–93.

Ruff, C. B., M. L. Burgess, R. A. Ketcham, and J. Kappelman. 2016. "Limb Bone Structural Proportions and Locomotor Behavior in A.L. 288-1 ('Lucy')." *Plos One* 11 (11): e0166095.

Ruxton, G. D., and D. M. Wilkinson. 2011. "Thermoregulation and Endurance Running in Extinct Hominins: Wheeler's Models Revisited." *Journal of Human Evolution* 61:169–75.

298

Rybczynski, W. 2000. *One Good Turn: A Natural History of the Screwdriver & the Screw*. New York: Simon & Schuster.

Samson, D. R., A. N. Crittenden, I. A. Mabulla, and A. Z. P. Mabulla. 2017. "The Evolution of Human Sleep: Technological and Cultural Innovation Associated with Sleep-Wake Regulation among Hadza Hunter-Gatherers." *Journal of Human Evolution* 113:91–102.

Samson, D. R., and C. L. Nunn. 2015. "Sleep Intensity and the Evolution of Human Cognition." *Evolutionary Anthropology* 24:225–37.

Samson, D. R., and W. R. Shumaker. 2015. "Orangutans (*Pongo spp.*) Have Deeper, More Efficient Sleep than Baboons (*Papio papio*) in Captivity." *American Journal of Physical Anthropology* 157:421–27.

Schallenberg, R. H. 1981. "Charcoal Iron: The Coal Mines of the Forest." In *Material Culture of the Wooden Age*, edited by B. Hindle, 271–99. Tarrytown, NY: Sleepy Hollow Press.

Sellers, W. I., G. M. Cain, W. Wang, and R. H. Crompton. 2005. "Stride Lengths, Speed and Energy Costs in Walking of *Australopithecus afarensis*: Using Evolutionary Robotics to Predict Locomotion of Early Human Ancestors." *Journal of the Royal Society Interface* 2:431–41.

Slater, D., R. S. Bradley, P. J. Withers, and A. R. Ennos. 2014. "The Anatomy and Grain Pattern in Forks of Hazel (*Corylus avellana L.*) and Other Tree Species." *Trees* 28:1437–48.

Smith, D. C. 1964. "Wood Pulp and Newspapers, 1867–1900." *Business History Review* 38:328–45.

Stephan, H., H. Frahm, and G. Baron. 1981. "New and Revised Data on Volumes of Brain Structures in Insectivores and Primates." *Folia Primatologica* 35:1–29.

Taylor, M. 1998. "Wood and Bark from the Enclosure Ditch." In *Excavations at a Neolithic Causewayed Enclosure near Maxey, Cambridgeshire, 1982–7*, edited by F. Pryor, 115–60. Swindon, UK: English Heritage.

Tegel, W., R. Elburg, D. Hakelberg, H. Stauble, and U. Buntgen. 2012. "Early Neolithic Water Wells Reveal the World's Oldest Wood Architecture." *Plos One* 7:e51374.

Thieme, H. 1997. "Lower Palaeolithic Hunting Spears from Germany." *Nature* 385:807–10.

Thorpe, S. K. S., R. L. Holder, and R. H. Crompton. 2007a. "Origin of Human Bipedalism as an Adaptation for Locomotion on Flexible Branches." *Science* 316:1328–31.

Thorpe, S. K. S., R. H. Crompton, and R. McN. Alexander. 2007b. "Orangutans Use Compliant Branches to Lower the Energetic Cost of Locomotion." *Biology Letters* 3:253–56.

Tree, I. 2017. *Wilding*. London: Picador.

REFERENCES

Turnbull, C. 1961. *The Forest People*. London: Jonathan Cape.

Uglow, J. 2002. *The Lunar Men*. London: Faber and Faber.

van Casteren, A., W. Sellers, S. Thorpe, S. Coward, R. Crompton, and A. R. Ennos. 2012a. "Why Don't Branches Snap? The Mechanics of Bending Failure in Three Temperate Angiosperm Trees." *Trees: Structure and Function* 26:789–97.

van Casteren, A., W. Sellers, S. Thorpe, S. Coward, R. Crompton, J. P. Myatt, and A. R. Ennos. 2012b. "Nest Building Orangutans Demonstrate Engineering Know-How to Produce Safe, Comfortable Beds." *Proceedings of the National Academy of Sciences* 109:6873–77.

van Casteren, A., W. Sellers, S. Thorpe, S. Coward, R. Crompton, and A. R. Ennos. 2013. "Factors Affecting the Compliance and Sway Properties of Tree Branches Used by the Sumatran Orangutan (*Pongo abelii*)." *Plos One* 8:7.

van der Woude, A., A. Hayami, and J. de Vries. 1990. *Urbanisation in History: A Process of Dynamic Interactions*. Oxford: Oxford University Press.

Vandiver, P. B., O. Soffer, B. Klima, and J. Svoboda. 1989. "The Origins of Ceramic Technology at Dolní Věstonice, Czechoslovakia." *Science* 246:1002–8.

van Schaik, C. P. 2004. *Among Orangutans: Red Apes and the Rise of Human Culture*. Cambridge, MA: Harvard University Press.

van Schaik, C. P., and C. D. Knott. 2001. "Geographic Variation in Tool Use on *Neesia* Fruits in Orangutans." *American Journal of Physical Anthropology* 114:331–42.

Vincent, A. S. 1984. "Plant Foods in Savanna Environments: A Preliminary Report of Tubers Eaten by the Hadza of Northern Tanzania." *World Archaeology* 17:131–47.

Waddington, C. 2007. *Mesolithic Settlement in the North Sea Basin. A Case Study from Howick, North-East England*. Oxford: Oxbow Books.

Waguespack, N. M., T. A. Surovell, A. Denoyer, A. Dallow, A. Savage, J. Hyneman, and D. Tapster. 2009. "Making a Point: Wood- versus Stone-Tipped Projectiles." *Antiquity* 83:786–800.

Warde, P. 2007. *Energy Consumption in England and Wales, 1560–2000*. Rome: Consiglio Nazionale Delle Ricerche.

Warman, P. H., and A. R. Ennos. 2009. "Fingerprints Are Unlikely to Increase the Friction of Primate Finger Pads." *Journal of Experimental Biology* 212:2015–21.

Warren, S. H. 1911. "On a Palaeolithic (?) Wooden Spear." *Quarterly Journal of the Geological Society of London* 67:xciv.

Westcott, D. 1999. *Primitive Technology: A Book of Earth Skills*. Salt Lake City: Gibbs-Smith.

REFERENCES

Wheeler, P. E. 1992. "The Influence of the Loss of Functional Body Hair on the Eater Budgets of Early Hominids." *Journal of Human Evolution* 23:379–88.

White, J. H. Jr. 1981. "Railroads: Wood to Burn." In *Material Culture of the Wooden Age*, edited by B. Hindle, 184–226. Tarrytown, NY: Sleepy Hollow Press.

Whitehouse, D. 2012. *Glass: A Short History*. London: British Museum Press.

Williams, M. 2002. *Deforesting the Earth: From Prehistory to Global Crisis*. Chicago: University of Chicago Press.

Winchester, S. 2018. *Exactly: How Precision Engineers Created the Modern World*. London: William Collins.

Woods, M., and M. B. Woods. 2000. *Ancient Construction: From Tents to Towers (Ancient Technology)*. Minneapolis, MN: Twenty-First Century Books.

Wrangham, R. 2009. *Catching Fire: How Cooking Made Us Human*. London: Profile Books.

Wrigley, E. A. 2010. *Energy and the English Industrial Revolution*. Cambridge: Cambridge University Press.

Yasuda, Y. 2012. *Water Civilization: From Yangtze to Khmer Civilizations*. Berlin: Springer Science & Business Media.

Yerkes, R. W., and B. H. Koldehoff. 2018. "New Tools, New Human Niches: The Significance of the Dalton Adze and the Origin of Heavy Duty Woodworking in the Middle Mississippi Valley of North America." *Journal of Anthropological Archaeology* 50:69–84.

Yorke, T. 2010. *Timber Framed Buildings Explained*. Newbury, UK: Countryside Books.

Zhou, H., J. Leng, M. Zhou, Q. Chun, M. F. Hassanein, and F. Wenzhou. 2018. "China's Unique Woven Timber Arch Bridges." *Civil Engineering* 171:1–21.

Index

Page numbers in italics refer to illustrations

About the Author

Roland Ennos is a visiting professor of biological sciences at the University of Hull. He is the author of textbooks on plants, biomechanics, and statistics, and his popular book *Trees*, which was published by London's Natural History Museum. He is an enthusiast for natural history, archaeology, and early music, and lives with his partner and several hundred ferns near Hull, in East Yorkshire, England.